Plant Proteins: Applications, Biological Effects, and Chemistry

ACS SYMPOSIUM SERIES **312**

Plant Proteins: Applications, Biological Effects, and Chemistry

Robert L. Ory, EDITOR
U.S. Department of Agriculture

Developed from a symposium sponsored by
the Division of Agricultural and Food Chemistry
at the 190th Meeting
of the American Chemical Society,
Chicago, Illinois,
September 8–13, 1985

American Chemical Society, Washington, DC 1986

Library of Congress Cataloging-in-Publication Data

Plant proteins.
(ACS symposium series; 312)

"Developed from a symposium sponsored by the
Division of Agricultural and Food Chemistry at the
190th Meeting of the American Chemical Society,
Chicago, Illinois, September 8–13, 1985."

Includes bibliographies and indexes.

1. Plant proteins as food—Congresses. 2. Food—
Protein content—Congresses. 3. Diet—Congresses.
4. Proteins in human nutrition.

I. Ory, Robert L., 1925–   . II. American Chemical
Society. Division of Agricultural and Food Chemistry.
III. American Chemical Society. Meeting (190th: 1985:
Chicago, Ill.) IV. Series.

TX392.A172   1986   641.1′2   86–10848
ISBN 0–8412–0976–6

ACS Symposium Series

M. Joan Comstock, *Series Editor*

Advisory Board

FOREWORD

The ACS SYMPOSIUM SERIES was founded in 1974 to provide a medium for publishing symposia quickly in book form. The format of the Series parallels that of the continuing ADVANCES IN CHEMISTRY SERIES except that, in order to save time, the papers are not typeset but are reproduced as they are submitted by the authors in camera-ready form. Papers are reviewed under the supervision of the Editors with the assistance of the Series Advisory Board and are selected to maintain the integrity of the symposia; however, verbatim reproductions of previously published papers are not accepted. Both reviews and reports of research are acceptable, because symposia may embrace both types of presentation.

CONTENTS

INDEXES

PREFACE

GROWING WORLD POPULATIONS, different degrees of income, religious beliefs, and health concerns have contributed to the increased interest in research on novel sources of protein from seeds, plants, and single-cell organisms. Such research has given rise to changes in consumption patterns and nutritional value of new protein food products. In addition, it has led to concern over the effects of potential antinutrients and large quantities of vegetable protein on health. Research on these subjects has been reported in several books (cf. Chapter 1, *Literature Cited*), but emphasis was devoted to the major cereals, oilseeds, and legumes (e.g., corn, wheat, barley, soybeans, peanuts, rapeseed). As technology has improved, the literature reporting new achievements in research on vegetable proteins has also grown.

Plans for the symposium from which this book was developed were initiated in 1982–83. The aim was to bring together specialists in three primary phases of research on plant proteins and to feature applications in both new and traditional foods, the biological effects of all-vegetable protein diets on human health, and the chemistry of some lesser known plant proteins that are important sources of food in some countries. The speakers and subjects were carefully selected to cover these three areas—the ABCs of plant proteins—and, although it was not possible to include all possible facets of these areas, the organizers sought to provide a good balance on important subjects covered very little or not at all in other symposia and books.

The problem of protein malnutrition is very complex and cannot be completely covered in one book, but the chapters of this book provide up-to-date balanced coverage of this important subject. This book, I hope, will fill a need for nutritionists, food scientists, and clinicians concerned about nutritional aspects of plant proteins for humans. Views and conclusions expressed herein are those of the authors, whom I sincerely thank for their time and effort in presenting their research at the symposium and in preparing their manuscripts for publication in this book.

ROBERT L. ORY
U.S. Department of Agriculture
New Orleans, LA 70179

December 10, 1985

Plant Proteins: The ABCs

Robert L. Ory

Southern Regional Research Center, Agricultural Research Service, U.S. Department of Agriculture, New Orleans, LA 70179

> What has been is what will be
> and there is nothing new under the sun.
> [Ecclesiates 1:9]

Interest in seed and vegetable proteins has been growing steadily over the past two decades because of the major role plant proteins play in both human and animal diets. Animal proteins are still acknowledged to have higher nutritional value than those from plant sources but for economic, health, or religious reasons, some populations derive all of their protein from plants.

In addition to this interest in new sources of protein, we have also seen a growing concern over antinutrients present in some plant foods and their effect on human health (1-3). Yet none of the plant, seed, or animal proteins we eat today are "new". They have been eaten for centuries - but in different forms. The "new" aspects include more modern ways of food preparation, better technology for measuring chemical composition, nutrients and antinutrients in foods, and a greater awareness of the health implications of food constituents.

Proteins are a vital part of living muscle tissue and are one of our most important nutrients. They have been called the building blocks of nutrition because they are broken down by digestive enzymes to provide amino acids for the building and repair of tissues. Animal proteins and those from most legumes and nuts contain all of the essential amino acids but the quantities of some (e.g.: lysine, methionine) in plants are lower. To envision the worldwide consumption of proteins in a different perspective, consider the major sources eaten by humans. Of the average 69 g. protein/day consumed worldwide in 1974, 63% was derived from plants (48% from cereals, roots and tubers, and 15% from fruits, vegetables, nuts, oilseeds and pulses), 36% from animal sources (meats, fish, eggs and dairy products), and 1% from other sources (4).

To achieve sufficient essential amino acids or a better amino acid balance (chemical score), plant proteins must be consumed in larger quantities or be blended with other complimentary proteins.

For example, peanut protein meal, low in lysine and methionine, can be blended with such things as high methionine citrus seed meal (5) or rice bran flour (6) to improve the chemical score. Use of less expensive plant proteins blended with complementary proteins is the logical and economical way to improve protein nutrition for those not consuming any animal protein, but this has raised questions concerning nutritional quality, presence of antinutrients, and physiological significance of such foods on human health. Major changes in life styles in the past two decades have also had an impact on eating habits; i.e.: more women work out today, life is lived at a more rapid pace, attitudes about meals prepared at home, meals eaten away from home, and members of families that eat together have changed. These changes have stimulated the introduction of many convenience foods, snack items, and interest in nutritional composition and labelling of processed foods.

In addition, consumption of fresh fruits and vegetables, milk, milk products, and red meats decreased during this period while consumption of snack items and fast food service products increased. Though fast food services are still the largest segment of the food-service industry, with a 3.5% increase in real growth between 1983 and 1984 (7), a growing awareness of nutrition by consumers is showing a slight trend back to fresh foods versus processed foods by some groups. Growth of snacks and convenience foods has, nevertheless, stimulated research on improving quality, flavor, color, texture, nutritional value and safety of these new food items. Research on plant and seed proteins has moved fast as technology improved and the literature reporting these achievements has also grown. In addition to hundreds of papers published in technical journals, there have been several recent books devoted to seed and plant proteins for human consumption. These were devoted to world protein supplies, functional and nutritional properties (8); chemistry, biochemistry and genetics of plant proteins (9, 10); or to structure, localization, evolution, biosynthesis, degradation, and improvement by breeding of seed proteins (11). Each of these provides excellent coverage of their subject matter but none focus on applications of plant proteins in traditional and new foods and on biological effects of plant proteins in the human diet. Plant foods are biologically more complex than animal food and, as noted earlier, plant proteins are nutritionally not as complete as animal proteins. Plants consist of many different tissues and structural elements that include fruits, seeds and seed hulls, nuts, roots and tubers, flowers, leaves, and stems. Some of these tissues are considered unsuitable for humans because of wide variations in nutritional value and/or inability to digest in the gastrointestinal tract and they serve as feed for animals. In contrast, animal foods are derived primarily from one tissue-muscle (plus eggs, milk and some organ meats), but far fewer species of animals serve as food sources for humans than do plant species.

The chapters in this book were carefully selected to complement the existing information on plant proteins by focusing on the A, B, C's: applications in new and traditional foods, biological effects of all-vegetable protein diets on humans, and composition and chemistry of some lesser-known sources of protein that could play an important role in areas where the major seed proteins (e.g.: soy, cottonseed, peanut, sunflower, rapeseed) may not be available.

Despite the wealth of information available on the biochemistry, genetics, and nutritional values of plant proteins, people eat foods that look, smell, and taste good; not because of nutritional importance. Thus, new blended plant foods or protein-supplemented snacks or food products will have to look and taste like the traditional items if they are to gain sufficient acceptance to become commercially feasible. Absolute food deficits are not the sole cause of hunger in the world since, theoretically, the 1.088 billion metric tons of food grains produced worldwide contain more than enough calories to provide the minimal requirement of 2,500 calories/day for its 3.5 billion people. The big problem is the uneven distribution of these resources. Not just in poor countries but even in small regions of the rich countries pockets of malnutrition still exist. Some reasons for the poor distribution, besides transportation problems, are the use of 27-30% of the grain for animal feed and almost a fourth of the supply is lost to insects, rodents, pathogens and waste.

Since most snack foods are based on cereals (wheat, corn, rice), a great deal of attention has focused on fortification/supplementation of traditional cereal-based foods. Worldwide, cereals represent the major source of calories and proteins for humans; i.e.: 52% and 47% of the world's average per capita intake of calories and protein, respectively (12). Cereal grains account for about 20% of the caloric intake in the U. S. but provide 55% in Mexico, 63% in India, and 67% in East Africa. Cereals provide 44% of the protein requirement in Mexico, 58% in Thailand, and 83% in the Middle East. The principal cereal of Latin America is maize (corn), in Asia it is rice, in the Middle East it is wheat, and in Africa, other grains such as sorghum.

Applications and uses of high protein legume flours in fortification of fried and baked goods and other food products for both Western and traditional diets of developing countries are covered in greater detail in Chapters 2-6. To achieve the balance needed in a treatise on food proteins and to include information on another growing use of vegetable proteins, that of "meat extenders" in Western diets, Chapters 7 and 8 describe the addition of plant proteins to processed meat products and whole muscle meats. The incorporation of plant protein (primarily soybean high protein meals and isolates) met with little success in the late 1960's/early 1970's because of flavor problems in defatted flours or meals. As technology improved, off-flavors were removed by production of concentrates and isolates, so that soy protein-extended ground beef products are used extensively today in school lunch programs, in military installations, and in several commercial ground beef products, hamburger, chicken and tuna helpers.

Biological effects of plant proteins on human health have attracted wide attention in the recent past because of the presence of various antinutrients such as trypsin inhibitors, hemagglutinins, and toxic principles (1). Adequate cooking and/or processing inactivates these materials and can improve the quality of plant foods, but the problem of sufficient essential amino acids in the protein is not resolved solely by cooking or processing. Blending of two or more proteins is still necessary to improve the chemical score and there are some biological effects caused by the protein quality and quantity. The effects of protein on skeletal integrity in early life and trace minerals utilization by rats and humans are discussed in Chapters 9 and 10. Additional chapters describing the effects of protein-procyanidin interactions on nutritional quality (Chapter 11), the acceptability of cottonseed protein foods by Haitian children (Chapter 12), effects of protein on experimental atherosclerosis (Chapter 13), and on intake and relation to cancer incidence in Seventh Day Adventists (Chapter 14) provide additional information on biological effects related to plant protein intake. Chapters 15 and 16 describe the effects of wet and dry processing on properties of legumes for food applications.

Because composition and nutritional properties of the major food legumes and oilseeds have been reported in numerous technical journals and books (listed above), the section devoted to composition and chemistry highlights lesser-known but potentially important sources of plant protein that have not received the same attention. Some of these food crops have been cultivated for many years so that they are not "new" sources. Such crops as winged bean, sweet potato, tropical seeds, fruits and leaves, yams and cucurbits are potential sources of protein in areas where they are grown. These are discussed in greater detail in the remaining five chapters.

The problem of protein malnutrition is too complex to be resolved with one single approach or single food. Nontechnical factors such as supply, availability, distribution, seasonal variations, age and health status of the consumer also play a role. However, with the technological advances made in the food industry today, we can now produce food products that are more nutritious and often cheaper than the traditional foods. Cereal grains, the world's principal source of food calories, can be fortified or supplemented with various plant proteins to produce very nutritious foods that look, taste, smell, and feel like traditional foods. The purpose here is certainly not to imply that animal products are bad for us but to show that there are also good proteins in plants that should not be overlooked in this search for edible proteins. It seems ironic that man has successfully conquered space by putting men on the moon and circling the earth in space stations but still has not eradicated hunger and malnutrition on earth.

Literature Cited

1. Ory, R. L. "Antinutrients and Natural Toxicants in Foods". Food and Nutrition Press, Inc., Westport, Conn., 1967.
2. Finley, J. W.; Schwass, D. E. "Xenobiotics in Foods and Feeds". American Chemical Society, Washington, D. C., 1983.
3. Ames, B. N. Science 1983, 221, 1256.
4. Dunne, C. P. J. Chem. Educ. 1984, 61, 271.
5. Ory, R. L.; Conkerton, E. J.; Sekul, A. A. Peanut Sci. 1978, 5, 31.

6. Conkerton, E J.; Ory, R. L. Peanut Sci. 1983, 10, 56.
7. Ellis, R. F. Food Process. 1984, 45 (9), 34.
8. Bodwell, C. E.; Petit, L. "Plant Proteins for Human Food".
    Martinus Nijhoff/Dr. W. Junk Publishers, The Netherlands,
    1983.
9. Harborne, J. B.; Van Sumere, C. F. "The Chemistry and
    Biochemistry of Plant Proteins". Academic Press, Inc.,
    London, 1975.
10. Gottschalk, W.; Muller, H. P. "Seed Proteins: Biochemistry,
    Genetics, Nutritial Value". Martinus Nihhoff/Dr. W. Junk
    Publishers, The Netherlands, 1983.
11. Daussant, J.; Mosse J.; Vaughan, J. "Seed Proteins".
    Academic Press, Inc., London, 1983.
12. Austin, J. E. Cereal Foods World 1978, 23(5), 229.

RECEIVED February 10, 1986

APPLICATIONS

2

Use of Peanut and Cowpea Flours in Selected Fried and Baked Foods

Kay H. McWatters

Department of Food Science, University of Georgia, College of Agriculture, Experiment Station, Experiment, GA 30212

Successful performance of legume flours as food ingredients depends upon the functional characteristics and sensory qualities they impart to the end product. In snack-type chips, cookies, and doughnuts, sufficient viscoelasticity should be provided to maintain product integrity, allow expansion, and develop surface character. Peanut flour possesses sufficient cohesiveness to form a fried snack-type product if low to moderate temperatures are used in processing the peanuts for flour production. Peanut and cowpea flours can replace at least 30% of the wheat flour in sugar cookies without altering extensively the diameter, height, spread, textural quality, and sensory attributes of the baked product. The quality of cake-type doughnuts containing peanut and cowpea ingredients was improved substantially by using the legumes in the form of flour rather than meal. Akara, deep-fat fried cowpea paste, may be prepared successfully from cowpea meal; conditions employed for pretreating seeds for mechanical decortication, meal particle size, and meal solids to water ratio were important considerations affecting functionality and product quality.

Successful performance of legumes in foods cooked by frying or baking indicate possible applications for their use. Poor performance, on the other hand, is useful in indicating areas where conditions of processing or preparation may need to be modified to accommodate legume flour usage. This presentation reviews work conducted in our laboratory to evaluate the performance of peanut and cowpea flours as ingredients in several food products.

Snack-Type Peanut Chips

Snack-type chips may be prepared by a fairly simple process from either peanut meal or flour. To prepare full-fat meal, peanuts are

0097-6156/86/0312-0008$06.00/0

heated at 93°C for 15 min in a forced draft oven, blanched to remove
the testa, coarsely ground in a food chopper, and passed through an
8-mesh screen to remove large particles. Water is then added to
form a dough-like mixture which may be shaped by forcing the dough
through a die or by rolling and cutting, followed by deep-fat
frying. This product form may be consumed alone or in combination
with dips or spreads.

Initial work to establish chip preparation conditions showed
that end product characteristics were influenced by meal particle
size, by the amount of water added to form the dough, and by the
length of time the dough was mixed (1). A very acceptable product
was achieved with these process conditions: a blend of particle
sizes most of which were in the 14-30 mesh range, an 18% added water
level, and a mixing time of 5 min. The final product had a crisp
texture, a typical roasted peanut flavor, and was quite similar in
composition to full-fat roasted peanuts. Chips contained about 49%
oil, 27% protein, and 1% moisture.

Commercial peanut flours were also evaluated for their
performance in preparation of peanut chips (2). The following
processing treatments were employed:

Flour Code	Processing Treatment
A	Full-fat, spray dried
B	Partially defatted, unroasted
C	Partially defatted, roasted at 160°C for 15 min
D	Partially defatted, roasted at 171°C for 15 min
E	Partially defatted, roasted at 177°C for 15 min

The processing conditions employed in preparation of the flours
influenced their appearance and color, dough-forming and handling
characteristics, frying time, and end product quality (Figure 1).
Flour A (full-fat, spray dried) produced a very oily, sticky dough
which was difficult to roll, shape, and handle. These chips required
a 2-min frying time to develop a golden brown color, became distorted
in shape during frying, and were extremely fragile to handle after
frying. Flour B (partially defatted, unroasted) lacked the oiliness
associated with flour A but was also difficult to handle because of
its sticky nature. Chips prepared from flour B required a 7-min
frying time to become golden brown in color and puffed slightly
during frying. Flour C (partially defatted, 160°C roast) produced a
dough that was easy to form and handle. Chips from flour C required
a 4-min frying time to develop a golden brown color and retained
their shape and structure during frying. Flour D produced a dough
that was neither sticky nor tacky but lacked cohesiveness. Chips
prepared from this flour completely disintegrated during the first
few seconds of frying, indicating that roasting conditions employed
in flour preparation had completely destroyed the flour's
cohesiveness. No attempts were made to prepare chips from flour E
since the temperature for roasting the nuts was even more severe
than that used for flour D.

Gel electrophoretic patterns of water-soluble proteins in the
five peanut flours were determined as previously described (2) and
show considerable differences in protein character (Figure 2). In

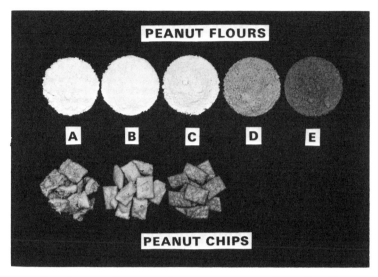

Figure 1.   Peanut flours and chips prepared from the flours.
Reproduced with permission from Ref. 2.   Copyright 1980,
Institute of Food Technologists.

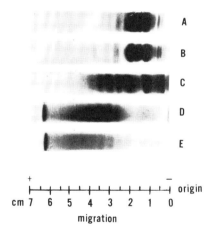

Figure 2.   Typical disc polyacrylamide gel electrophoretic
patterns of water-soluble proteins from peanut flours.
Reproduced with permission from Ref. 2.   Copyright 1980,
Institute of Food Technologists.

flours A and B, the components of the major globulin, arachin, are
in the 0.5 to 2.0 cm region and the nonarachin proteins appear in
the 2.0 to 4.0 cm region. The process conditions employed for flour
C altered some of the arachin components (region 1.3 to 2.0 cm) and
the nonarachin proteins to forms where they moved as diffuse bands
in the gel patterns. The process conditions used for flours D and E
denatured virtually all of the proteins to small polypeptides and/or
aggregates; this was accompanied by a loss of binding capacity and
cohesiveness in chip preparation.

Legume Flour Sugar Cookies

Sugar cookies may be prepared by replacing a portion of the wheat
flour with non-wheat flours derived from peanut, soybean, and cowpea
(3). The legume flours used in this study were adjusted to a
uniform oil content of about 1% by solvent extraction. Flour
protein content (wet weight basis) was 51% for peanut, 46% for
soybean, and 21% for cowpea. The flours were used at 0, 10, 20, and
30% wheat flour replacement levels. Peanut and cowpea flours could
replace at least 30% of the wheat flour without altering extensively
the diameter, height, spread, top grain, textural quality, and
sensory attributes of the baked product (Figure 3). Therefore, no
alterations in formulation or preparation procedures would be
required to accommodate the use of peanut or cowpea flours in this
type of cookie. Use of soybean flour at the 20 and 30% levels,
however, decreased cookie diameter and spread ratio, increased
cookie height and hardness, and prevented development of a typical
top grain during baking. Peanut flour cookies received high sensory
scores for all attributes. Cowpea flour products were also highly
acceptable except at the 30% level where a beany aroma and flavor
were noted. The soy flour cookies at the 10% level were quite
acceptable, but the poor spread of the 20 and 30% soy products
adversely affected texture quality. A beany flavor was also
apparent at the higher soy levels. In subsequent studies, soy flour
cookies with good spread characteristics were produced by increasing
the amount of water in the formula to compensate for soy flour's
high water absorption capacity.

The protein content of cookies was markedly influenced by the
addition and protein content of the various legume flours (Figure 4).
Each increment of peanut flour raised the total protein content in
cookies by 1.5%. Increases of 1.4% occurred with soy flour and 0.5%
with cowpea flour.

Legume Flour Doughnuts

The performance of peanut and cowpea flour as ingredients in
cake-type doughnuts has been investigated (4). In addition to
sensory characteristics imparted to the end product, the influence
of the legume flours on the machinability of the doughnut batter was
found to be important. Machinability is defined as the ease of
cutting, dispensing, and conveying the batter. Initial tests
indicated that peanut and cowpea meals were compatible ingredients
for use in this type of product, but the fat content of
legume-supplemented doughnuts was considerably higher than reference
doughnuts made with 100% wheat flour.

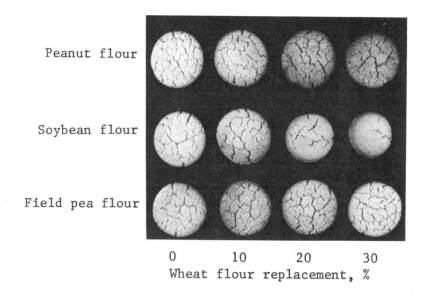

Peanut flour

Soybean flour

Field pea flour

0   10   20   30
Wheat flour replacement, %

Figure 3. Representative sugar cookies prepared from defatted
peanut, soybean, and field pea flours at 0, 10, 20, and 30%
wheat flour replacement levels. Reproduced with permission from
Ref. 3. Copyright 1978, American Association of Cereal Chemists.

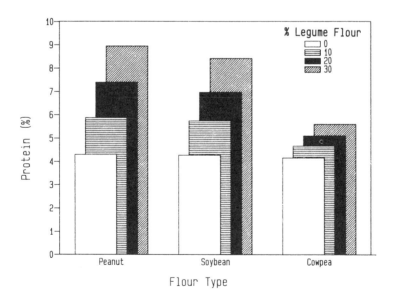

Figure 4. Protein content (%) of sugar cookies prepared from
defatted peanut, soybean, and cowpea flours at 0, 10, 20, and
30% wheat flour replacement levels. Reproduced with permission
from Ref. 3. Copyright 1978, American Association of Cereal
Chemists.

Follow-up studies utilized finely-milled legume flours and the addition of soybean flour as a fat-control agent in an effort to improve doughnut quality (5). The legume products and doughnuts prepared from them are shown in Figure 5. On a dry weight basis, peanut flour from solvent extracted peanuts (PF-SE) contained 0.9% fat and 54.4% protein while cowpea flour (CF) contained 1.4% fat and 25.5% protein. Peanut flour from partially defatted untoasted peanuts (PF-PD-U) contained 34.5% fat and 34.9% protein while peanut flour from partially defatted peanuts toasted at 160°C contained 34.4% fat and 37.6% protein.

Wheat flour reference (WFR) samples were compared to doughnuts made with 10% legume flour (LF) or legume flour plus soybean flour (LF + S) added to increase the total flour content by 3%. No machinability problems were encountered except with peanut flour from partially defatted, untoasted peanuts. This flour produced a sticky batter which prevented the cutting/dispensing device from completely cutting away the doughnut center. All of the legume flour products received very acceptable sensory ratings. Panelists were able to detect a "slightly beany" aroma in products made with the partially defatted, untoasted peanut flour and a "definite roasted peanut" aroma in doughnuts made with partially defatted, toasted peanut flour. The grain of legume flour-supplemented doughnuts was uniform and fine-textured, and the fat content of the test products was acceptable, closely resembling that of reference doughnuts. The addition of soy flour provided no added benefit in controlling the amount of fat absorbed by the doughnuts during frying.

Akara (Fried Cowpea Paste)

The University of Georgia is participating in a collaborative research project with the University of Nigeria. The overall goal of the project is to increase the availability of cowpeas for developing country populations by development of technologies to reduce postharvest storage losses and to simplify the preparation of cowpeas. The project is supported by the Bean/Cowpea Collaborative Research Support Program (CRSP) and the U. S. Agency for International Development.

Cowpeas are an important part of the diets of West Africans. They're grown for domestic consumption rather than for export and are consumed as a boiled vegetable and as paste which is cooked by either steaming or frying.

In order to obtain a light-colored paste free of any color which may be imparted by the seed coat or eye, peas are decorticated by manual processes. The traditional wet method consists of soaking peas in water to loosen the seed coat, then rubbing the seeds to separate the testa from the cotyledon. The decorticated peas are then ground into paste either in a mortar, on a stone, or in a blender if available. The paste is stirred or whipped to incorporate air, seasoned, then cooked (6,7). Chopped peppers and onions are the most common seasonings, but tomatoes, ginger, or shrimp are also used for flavor. The fried product (Figure 6), called akara in Nigeria, is consumed as a snack food and breakfast food and has a bread-like character similar to hush puppies. In addition to home preparation, akara is prepared and sold by street vendors in the

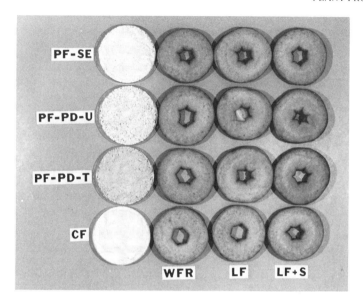

Figure 5. Representative doughnuts prepared from peanut flour-solvent extracted (PF-SE), peanut flour-partially defatted-untoasted (PF-PD-U), peanut flour-partially defatted-toasted (PF-PD-T), and cowpea flour (CF). WFR = wheat flour reference, LF = 10% legume flour (peanut or cowpea), LF + S = 10% legume flour (peanut or cowpea) + 3% soybean flour. Reproduced with permission from Ref. 5. Copyright 1982, The American Peanut Research and Education Society.

Figure 6. Representative batch of akara, a bread-like product made by deep-fat frying cowpea paste. Reproduced with permission from Ref. 6. Copyright 1980, Institute of Food Technologists.

marketplace and by small-scale processors for home delivery and
catering services.
    A major focus of the project has been development of a process
to make cowpea meal or flour that could be implemented at the
village level. The meal would be a ready-to-use convenience product
to which the consumer could simply add water and then proceed to the
final steps of preparation. The time-consuming and labor-intensive
steps of soaking, decorticating, and grinding would be eliminated.
If a convenience product such as cowpea meal is to find favor with
the consumer, it must produce an end product which is similar in
quality to that produced by the traditional process. Essential to
attainment of a light, spongy texture in the fried product is the
formation, during the whipping of cowpea paste, of a foam with
appropriate volume and consistency.
    Commercial cowpea flour available in Nigeria has not been well
received by consumers because of its poor water absorption and
because akara prepared from the flour is heavy, lacks crispness, and
lacks the flavor typical of products made from fresh paste (8).
A major difference found between traditional paste and commercial
flour obtained in Nigeria was particle size distribution (9). The
flour was more finely milled than traditional paste with 47% of the
flour particles riding a 400-mesh screen compared to 16% at the
400-mesh size for the paste (Figure 7). The greatest concentration
of paste particles (64%) was in the 50-100 mesh range whereas most
of the flour particles (68%) were concentrated in the 200-400 mesh
range. Attempts to make akara from the flour, using instructions
supplied by the manufacturer, were unsuccessful because the batter
was too liquid to dispense and fry properly. Even after reducing
the amount of water in the batter, cooked products were unattractive
and undesirably dry, dense, and tough. These findings prompted
further study on the effects of particle size on cowpea paste
characteristics and akara-making quality.
    Three screen sizes (2.0, 1.0, 0.5 mm) were used for milling
cowpeas and produced the particle size distributions shown in Figure
8. With the 2.0 mm screen, particles were concentrated (76%) in the
30-100 mesh range. With the 1.0 mm screen, most of the particles
(82%) were in the 50-200 mesh range. Eighty per cent of the
particles were in the 200-400 mesh range with the 0.5 mm screen.
    In preparing akara from each milled product, too many large
particles still remained in the 2 mm material to make a smooth paste.
However, highly acceptable akara with uniform shape was produced
from this material after the paste was ground to eliminate the large
particles. With the 0.5 mm screen, the paste was very fluid and
extremely difficult to dispense, behavior which closely resembled
that exhibited by the commercial cowpea flour. Akara prepared from
the 0.5 mm material was also extremely distorted. Of the three
screen sizes compared, the 1.0 mm screen produced the most desirable
particle size distribution; although the paste produced from the 1.0
mm material was somewhat more fluid than desired, it appeared that
adjustments could be made in hydration of the meal to achieve an
appropriate batter viscosity.
    The poor performance of finely milled cowpea flours may be due
to changes in physical form and structure which occur as a result of
milling. Sefa-Dedah and Stanley (10) investigated the relationship
of microstructure of cowpeas to water absorption and decortication

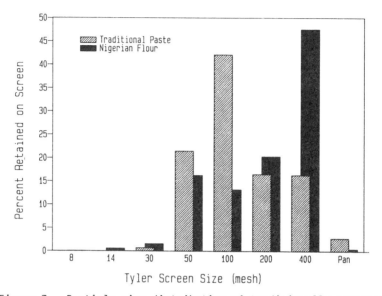

Figure 7. Particle size distribution of traditionally processed
cowpea paste and mechanically milled cowpea flour. Reproduced
with permission from Ref. 9. Copyright 1983, American
Association of Cereal Chemists.

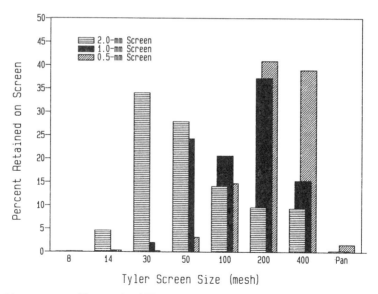

Figure 8. Effect of mill screen size on particle size
distribution of cowpea meal/flour. Reproduced with permission
from Ref. 9. Copyright 1983, American Association of Cereal
Chemists.

properties and found that water uptake of whole seeds was a
sequential process related to the seed's cellular structure. When
the form of the seed becomes so altered by a process such as milling
that structural and compositional components are vastly different
from those of the intact seed, then the conditions which allow for
the sequential uptake of water no longer exist.

In subsequent studies to determine the most appropriate water
level for hydrating cowpea meal produced from the 1.0 mm screen,
sufficient water was added to the meal to adjust the moisture
content to 56, 58, or 60% (11). Preliminary studies had shown that
a 54% water level produced a batter that was too thick for whipping,
dispensing, and frying, and a 62% water level was too thin.
Traditional paste made from soaked peas contains about 61% water and
has a viscosity value after whipping of about 302 poise. By
comparison, the viscosity of paste made from hydrated cowpea meal
was 578 poise at the 56% water level, 441 poise at the 58% water
level, and 333 poise at the 60% level. The 60% water level produced
paste with flow properties and a cooked product with physical
characteristics more like the traditional product than the other
water levels.

Sensory attributes of akara made from the 1 mm screen flour
hydrated to a 60% moisture content before cooking were acceptable
when compared to traditional akara (11). A major difference in
akara prepared from hydrated meal and that prepared from traditional
paste is in the fat content of the cooked product. On a dry weight
basis, traditional akara contains about 38% fat whereas akara made
from meal hydrated to a 60% moisture content contains 29% fat. A
frequent comment made by sensory panelists is that akara made from
meal has a drier texture and mouthfeel than traditional akara.
Since the moisture content of traditional and meal-based akara is
similar (about 45%), the perceived drier texture of the meal
products is probably due to their lower fat content.

Although a fried cowpea paste product such as akara is
unfamiliar to consumers in the Western world, this use for cowpeas
may have application as a snack food or as a bread-like
accompaniment for fish or poultry. Legumes already play an
important role in the diets of the world's population. Applications
in which legumes perform successfully increase the potential for
extending their usage even further.

Acknowledgments

Supported by State and Hatch funds allocated to the Georgia
Agricultural Experiment Stations, by a grant from the Bean/Cowpea
Collaborative Research Support Program (U. S. Agency for
International Development), and by funds from the Georgia
Agricultural Commodity Commission for Peanuts.

Literature Cited

1. McWatters, K.; Heaton, E. K. Univ. Georgia Coll. Agric.
   Experiment Stations Research Bulletin 106, 1972; pp. 1-17.
2. McWatters, K. H.; Cherry, J. P. J. Food Sci. 1980, 45, 831.
3. McWatters, K. H. Cereal Chem. 1978, 55, 853.
4. McWatters, K. H. Peanut Sci. 1982, 9, 46.

5. McWatters, K. H. Peanut Sci. 1982, 9, 101.
6. McWatters, K. H.; Flora, F. Food Technol. 1980, 34 (11), 71.
7. McWatters, K. H.; Brantley, B. B. Food Technol. 1982, 36 (1), 66.
8. Dovlo, F. E.; Williams, C. E.; Zoaka, L. International Develop. Research Center Public. IDRC-055e, 1976; pp. 1-96.
9. McWatters, K. H. Cereal Chem. 1983, 60, 333.
10. Sefa-Dedah, S.; Stanley, D. W. Cereal Chem. 1979, 56, 379.
11. McWatters, K. H.; Chhinnan, M. S. J. Food Sci. 1985, 50, 444.

RECEIVED December 13, 1985

Use of Field-Pea Flours as Protein Supplements in Foods

Barbara P. Klein and Martha A. Raidl[1]

Department of Foods and Nutrition, University of Illinois at Urbana–Champaign, Urbana, IL 61801

Legume seeds, such as soy and other pulses, are widely used as protein sources in the human diet. Recent advances in technology suggest that protein concentrates and isolates made by relatively simple methods can be incorporated into food products. Flours made from field peas by wet or dry milling, or air classification, possess distinctive sensory, functional and nutritional characteristics. Compositional differences in the pea seeds influence the quality of the end products. Pea flours have been used for protein enrichment of a number of cereal-based products; however, undesirable sensory characteristics may limit their use, in spite of improved functional effects in food systems. The production of volatile compounds during cooking and baking of foods with pea supplementation affects their acceptability. Enzyme systems active in unheated pea flours may contribute to their functional properties, but adversely affect the sensory quality of the food.

Legume seeds, such as soybeans (Glycine max L.), faba beans (Vicia faba L.), cow peas (Vigna unguiculata L.), navy beans (Phaseolus vulgaris L.) and field peas (Pisum sativum L.), are widely used as protein sources in the human diet. In many parts of the world, legumes are a major contributor to both caloric and protein intakes. The advantages of using beans are many: they have long storage lives, even under adverse environmental conditions; they are easily transported; and they require minimum equipment for preparation.

Soybeans have received more attention than any of the other legumes or pulses. Their high protein and oil content make them a valuable commodity, both from an economic and nutritional standpoint. The major uses of soybeans have been in processed

---

[1] Current address: Coca-Cola Foods, P.O. Box 550, Plymouth, FL 32768.

0097–6156/86/0312–0019$06.00/0
© 1986 American Chemical Society

products, either as oil or as high protein flours and concentrates. The functionality of soy products differs widely depending on the time, temperature and moisture conditions used in processing (1).

Legumes other than soy are more often consumed as the whole bean, either cooked or ground into a flour. Consumption of legumes is higher in less developed countries than in the more industrialized. In India, for example, consumption of legumes and pulses exceeds 60 grams per capita per day (2) while in the United States and the United Kingdom, it is only about 10 to 15 grams per capita per day (3).

Field peas, both yellow and green, are grown in Canada, the Northwest United States, and to a lesser extent, in Northern Europe. In the United States, the primary interest in peas has been in the fresh, immature, green vegetable, which is canned or frozen, rather than in the mature dry seed. Thus, much of the genetic research has been directed towards improving appearance, yield, disease resistance, canning and freezing quality, instead of attempting to increase protein content or quality. In the past ten years, advances in processing technology have made it possible to produce pea protein concentrates by some of the same methods used for soy proteins, and more importantly, by relatively simple methods such as air classification.

Variability in Composition of Field Peas

The composition of the field pea depends not only on the species, but also on the cultivar that is being processed (4,5). Variations exist among cultivars (e.g., Trapper, Century) in protein, fat, carbohydrate (crude fiber and starch), and ash contents, as shown in Table I. Tyler and Panchuk (6) noted that the composition of field peas at different stages of maturity also affected the composition of the products, and this could ultimately influence their functionality in foods.

Table I.   Proximate Composition of Peas (g/100 g)

	Protein	Fat	Total Carbohydrate		Ash
Dry seeds (7)	24.1	1.3	60.3		2.6
			Starch	Fiber	
Trapper cv. (8)	14.5	4.1	59.8	4.3	3.3
	18.3	3.7	56.7	3.7	3.0
	24.2	3.3	53.8	3.5	2.7
	28.5	3.0	49.7	3.1	2.8
Century cv. (4)	23.3	1.2	54	7.6	2.5

Protein Components of Field Pea Flour

The reported protein content of field peas ranges from 13.3 to 39.7%

and is influenced by genetic and environmental factors (8-13).
Analysis of 1452 varieties of field peas (Pisum sativum) showed that
the protein varied from 15.5 to 39.7% (10), and even genetically
identical pea plants grown the same year on the same field produced
seeds whose protein content ranged from 19.3 to 25.2% (11). More
than 75% of all field peas grown are of the Trapper variety and
their protein content can range from 13.3 to 27.1% (8). Watt and
Merrill (7) indicate that the protein content of mature dry peas is
24.1%, but Reichert and MacKenzie (8) found that only 14% of their
198 field pea samples had protein levels greater than or equal to
this value (see Table I).

Environmental factors which affect protein content of field
peas include nitrogen fertilizer (14), maturation (15), soil P and K
content (16), and temperature (17). The protein content of field
peas is important since it ultimately affects the amount of protein
in field pea concentrates or isolates (18).

Protein content of dehulled Trapper field peas is negatively
correlated with the amino acids threonine, cystine, glycine,
alanine, methionine, and lysine and positively correlated with
glutamic acid and arginine (8). Holt and Sosulski (19) obtained
similar correlations with Century field peas for all amino acids
except glutamic acid. Other investigators (20) also found that
sulfur amino acids (cys, met) are negatively correlated with protein
content.

The main storage proteins in field peas are two globulins
(Table II), vicilin and legumin, which are similar to the 7S and 11S

Table II. Proteins in Peas and Soybeans (g/100 g) (21)

	Albumins[a]	Globulins[b]	Glutelin[c]
Peas	21	66	12
Soy	10	90	0

[a]Soluble in water.
[b]Soluble in salt solution.
[c]Soluble in dilute acid or base.

fraction of soy protein (12). However, legumin appears to have a
more compact structure than the 11S soybean fraction, and vicilin,
although comparable to the 7S fraction, is thought to be two
proteins. Vicilin has a molecular weight of 186,000; legumin is
approximately 331,000. These proteins do not participate to the
same degree as soy proteins in association-dissociation reactions
when there is a change in ionic strength (21). The two pea
globulins differ in their properties: for example, vicilin is
soluble at pH 4.7, legumin is not; legumin is not heat-coagulable,
but vicilin is. The globulins in peas appear similar to those in
other legumes as well as in soybeans.

The protein efficiency ratio (PER) of field pea flours is considerably less than that of casein (1.46 vs. 2.50), and somewhat less than that of soy flour (1.81). However, composites of wheat flour and pea or rice and pea (50% of the protein from each source) had PER's of 2 or more (22). Thus, supplementation of cereals with pea flour results in improvement of protein quality.

Starch Components of Field Pea Flour

Starch content of field peas (Pisum sativum L., cv. Trapper) ranges from 43.7 to 48% and, after being subjected to pin milling and air classification, produces a flour containing 78% starch (9,12,13).
    The predominant polysaccharide in dehulled field pea flour is starch (49.7-59.8%) and the major soluble sugars are α-galactosides (4.78%) and sucrose (1.85-2.2%) (8,23,24). Verbascose is the major α-galactoside present in field pea flour (23,24). The α-galactosides are the main contributors to the flatulence caused by ingestion of legume flours.
    Pea starch granules are oval, sometimes fissured, with a diameter of 20-40 μm (13). Molecular and structural characteristics of the two main components of field pea starch--amylose and amylopectin--are important in determining functional properties (25,26). Smooth field pea starch concentrate contains 97.2% starch of which 30.3-37.8% is amylose (9,23,25-27), and wrinkled pea starch concentrate contains 94.8% starch, which is 64% amylose (26). The gelatinization temperature of smooth pea starch is between 64 to 69 C, and that of wrinkled pea starch is greater than 99 C to 115 C. Gelatinization temperature depends on maturity of field pea seed and amylose content (26,27).

Processing Methods for Pea Flours

The field pea seed is first cleaned, usually dehulled, and ground to a flour prior to being separated into starch and protein fractions (4). The flour is pale yellow or green, depending on the cultivar. The separation of field pea flour into protein and starch concentrates is achieved using either a wet or dry milling process. The wet processing of field peas produces a relatively pure protein concentrate or isolate composed of approximately 60% protein and a starch fraction containing about 2% protein (22). Unfortunately this process requires evaporation of large amounts of water, making it expensive and technologically complex (4).
    Dry processing of field peas uses pin milling and air classification techniques (4,23). Whole or dehulled field pea seeds are pin milled to yield flours with a specific particle size which can be further separated into protein and starch fractions using an air classifier (9). In this system, using an Alpine Air Classifier for example, a spiral flow of air is used to separate the jagged and "light" protein particles from the smooth, round and "heavy" starch granules, respectively, into fine and coarse fractions (4,18). The starch fraction is then washed, centrifuged, and defibered to yield a pure starch concentrate (9). Many investigators have found air classification effective in separating starch and protein-rich fractions in other starchy grain legumes as well as in field peas (4,18,23,28-30).

The composition of protein and starch fractions produced from pin milling and air classification are related to a number of variables: variability in composition of field pea cultivars, number of passes through pin mill and air classifier, vane settings and protein content of peas, and seed moisture (5,9,23,31).

Protein content of field peas is negatively correlated with lipid, cell wall material (CWM), sugar, and ash content and positively correlated with starch separation efficiency and protein separation efficiency in air classification of pea flour. The lower separation efficiency of low protein peas may be due to their high lipid and CWM content which makes disintegration of seeds and separation into protein and starch particles by pin milling difficult. It is suggested that peas with a specific protein content should be used in order to control the protein and starch fraction contents (18).

As seed moisture in field peas decreases, there is a decrease in starch fraction yield, protein content of starch fraction, protein content of protein fraction, and percent starch separation efficiency, and a concurrent increase in protein fraction yield, percent starch in starch fraction, percent starch in protein fraction, percent protein separation efficiency, and percent neutral detergent fiber in the protein fraction. Lower moisture content of field peas improves milling efficiency and results in more complete separation of protein and starch fractions, which could explain the increase in protein fraction yield and percent starch in starch fraction, improved protein separation efficiency and less protein in the starch fraction. The decrease in starch separation efficiency was probably due to the increased starch content of protein fraction and increased protein fraction yield with lower seed moisture. Finer grinding of CWM may explain the increase in NDF in protein fractions (32).

Pea Protein Concentrates and Isolates

Isoelectric precipitation and ultrafiltration procedures have been used to produce protein isolates from field peas (13). Sumner et al. (33) used an alkaline extraction method to produce pea protein isolate either as sodium proteinate or as an isoelectric product which was then dried using either a spray, drum, or freeze drying method. The isoelectric process and ultrafiltration process produced field pea protein isolates which contained 91.9% and 89.5% protein, respectively (13). The spray, freeze, and drum drying processes produced sodium proteinate isolates which contained 85.8, 83.0, and 83.2% protein, respectively, while their isoelectric counterparts contained 88.5, 90.0, and 85.9% protein (33). The resulting pea protein isolate is a cream to beige color and tastes fairly bland (13). The color depends on the method used to dry the isolates. Spray-dried isolates are the lightest, while freeze-dried and drum-dried are the darkest. Oxidation of polyphenols causes the darkening of freeze-dried products while the Maillard reaction from heat processing creates a darker product in drum-dried isolates (33).

Carbohydrate Content. Protein fractions were found to contain 40 to

90% higher levels of α-galactosides when compared to the corresponding field pea flours. Thus, the protein fraction contained high levels of verbascose and stachyose and the major galactosides remained with the protein fraction during air classification (24).

Amino Acid Content. Amino acid content of field pea products is related to protein level, method of processing, and fraction (starch or protein). The protein fraction contains fewer acidic (glu, asp) amino acids than the starch fraction and more basic (lys, his, arg) amino acids than the starch fraction. Also, there are more aromatic (tyr, phe) amino acids, leu, iso, ser, val, and pro in the protein fraction than in the starch fraction (5). An amino acid profile of pea protein concentrate shows relatively high lysine content (7.77 g aa/16 g N) but low sulfur amino acids (methionine and cystine) (1.08-2.4 g aa/16 g N). Therefore, it is recommended that air classification or ultrafiltration be used because acid precipitation results in a whey fraction which contains high levels of sulfur amino acids (12,23). Also, drum drying sodium proteinates decreases lysine content due to the Maillard reaction (33).

Nitrogen Solubility Index. Nitrogen solubility index (NSI) indicates the extent of denaturation of a protein and correlates well with the functional characteristics of protein ingredients. NSI values are influenced by a number of factors, such as pH, temperature, particle size of product, process used for protein isolation, and protein content (34).

Pea protein isolate produced at pH 3 and 7 using ultrafiltration exhibited 81% nitrogen solubility and only 66% solubility when the isoelectric precipitation method was used (13). Sodium proteinate isolates subjected to freeze, spray or drum drying processes had lower nitrogen solubility than the corresponding isoelectric protein isolates (33). The percent nitrogen solubility of the isolates varied from 0 to 100% over a pH range of 3-10. The lowest nitrogen solubility values occurred at pH 4.5 (the isoelectric point) for all products. The low NSI values for drum dried sodium proteinate over this pH range were probably due to protein denaturation during processing. Higher NSI values occurred in the relatively undenatured spray- and freeze-dried pea protein isolates (33).

Nitrogen solubility index is inversely related to protein level, i.e., as the protein level increases, NSI decreases (8). Another factor related to solubility of seed nitrogen in a flour and distilled water suspension is the concentration of water-soluble naturally occurring salts, since salt-soluble globulins are the major proteins found in peas (21). Also, differences in pea mineral content may play a role in NSI.

Water Absorption. Water absorption of pea protein isolates depends on pH and processing method used to produce the isolate. Isoelectric pea protein isolate absorbed 2.7 to 2.8 times its weight of water at pH 7 while UF pea protein isolate absorbed 3.3 times its weight of water at pH 2.5 and twice its weight in water at pH 8.5 (13). These low water absorption values may be due to the high nitrogen solubilities of these proteins (35).

Use of Field Pea Products in Cereal-Based Products

Although there are some references to the incorporation of pea flours or concentrates in meat systems (12,22,36), the primary use of field pea flours or concentrates has been in baked products or pastas. Most of the studies conducted on baked products have been with pea flour that has had little or no heat treatment. Thus, the flour can be considered enzyme-active. As a storage organ for the plant, the intact pea seed contains a complex assortment of enzymes, including amylases, proteases, lipases and lipoxygenases, as well as a variety of oxidative enzymes necessary for seed metabolism and germination. The function of the enzymes from the physiological standpoint in the plant is very different from their effects in a food system. Peas contain very low levels of anti-nutritional factors, namely trypsin inhibitor and hemagglutinins, when compared with soy (37). Thus, heating of pea flours is not essential to inactivate these compounds.

Field Pea Flours in Pasta. Incorporation of non-wheat flours into noodles improves the protein content and quality, but may have an adverse effect on the flavor and texture of the pasta. Hannigan (38) reported that 10% substitution of wheat flour with pea or soy flour resulted in satisfactory quality of Japanese Udon noodles. When the pea flour was heated, the flavor was considerably improved. Cooked yellow pea flour-fortified noodles were comparable to the control with respect to sensory characteristics and yield.
    Nielsen et al. (39) used pea flour and pea protein concentrate, both cooked and raw, in noodles and spaghetti. The pasta was made from composite flours prepared by blending 33% pea flour with 67% wheat flour or 20% pea concentrate with 80% wheat flour. Protein content of the fortified noodles was approximately one-third higher than the wheat flour noodles. Addition of pea flour reduced the cooking time, but resulted in a softer product and lower yield than the wheat pastas. Precooking the pea flour improved flavor and decreased noodle dough stickiness, but the texture and yield of the cooked pasta was still less than that of wheat products.

Field Pea Flours in Bread Products. Legume flours, particularly soy, have long been incorporated into wheat-based products, both for their functional effects and for protein fortification. In general, increasing the levels of legume flours results in decreased loaf volume, lower crumb grain quality, and adverse flavor characteristics in the baked bread (Table III).
    Results have varied with respect to the amount of field pea flour that can be incorporated into a yeast bread before an unacceptable product is produced. Tripathi and Daté (40) made breads containing 5, 10 and 15% field pea flour and found breads made with more than 5% pea flour were not acceptable. Loaf volume decreased as the percent substitution increased. At the 5% level, color, flavor and taste of the breads were rated as excellent, but at the 10 and 15% levels, there was a bitter taste.
    Fleming and Sosulski (45) found that the incorporation of field

pea concentrates into yeast breads at levels from 5 to 25% increased
the protein content from 10.2% for the wheat control to 16.8% at the
highest substitution level. However, specific volume of the loaf

Table III. Baked Products Made with Field Pea

Product	Levels of Substitution	Quality
Yeast bread (40)	5, 10, 15% pea flour	Decreased volume Acceptable sensory quality at 5% Bitter taste at 10 and 15%
Yeast bread (37)	5-20% raw or cooked pea flour	Decreased volume Bleaching effect with raw pea flour Lower acceptability above 15%
Yeast bread (41)	8 and 15% concentrate	Decreased volume Lower acceptability at 15%
Yeast bread (42)	2.5-10% pea flour	No volume change Acceptable, beany at 10%
Quick bread (43)	5-15% pea flour	No volume change Beany flavor at 10%
Biscuits (44)	8% pea flour	Aroma and flavor decreased Doughy texture

decreased sharply (from 6.04 to 3.56 cc/g) and crumb grain and loaf
shape scores were steadily and significantly decreased with each 5%
increment in soy flour. They observed similar results with other
plant protein concentrates. The addition of dough conditioners such
as sodium stearoyl lactylate improved the volumes and crumb grain
ratings of the breads. Sosulski and Fleming (41) also reported that
addition of 8 or 15% field pea flour plus 2% vital gluten resulted
in breads that were generally acceptable, particularly at the lower
level of substitution.

Jeffers et al. (32) used 5, 10, 15 and 20% substitutions of raw
and cooked pea flour in wheat bread. Different levels of KBrO₃
were incorporated in the doughs. Mixing times were decreased
significantly when pea flour was used. Mixing tolerance increased
at 5 and 10% levels, but was less at 15 and 20% levels with the raw
pea flour; cooked pea flour did not improve mixing tolerance. Loaf
volumes decreased with increasing levels of pea flour, as did crumb
grain scores. However, at the 15% substitution levels, the breads
were nearly standard.

Repetsky and Klein (42) found that pea flour significantly
affected the texture, color and flavor of yeast breads. At

substitution levels ranging from 2.5 to 10%, loaf volume and specific volume of the breads were not significantly different from the wheat flour control. But trained taste panelists detected a beany off-flavor in breads with 10% pea flour, and the color scores were lower than for the control.

The baking studies with yeast breads indicate that levels of substitution of up to 15% pea flour or concentrate for wheat flour result in breads that are generally acceptable, but are readily distinguishable from, and less preferred to, wheat controls. Although the use of a cooked or heated pea flour or concentrate improves flavor, the functionality of the product in bread is adversely affected.

Field Pea Flour in Other Baked Products. When McWatters (44) substituted 8% field pea flour and 4.6% field pea concentrate for milk protein (6%) in baking powder biscuits, sensory attributes, crumb color, and density of the resulting biscuits were adversely affected. No modifications were made in recipe formulation when pea products were incorporated. The doughs were slightly less sticky than control biscuits that contained whole milk. This might be due to lack of lactose or to the different water absorption properties of pea protein or starch. Panelists described the aroma and flavor of these biscuits as harsh, beany and strong. Steam heating the field pea flour improved the sensory evaluation scores, but they were never equivalent to those for the controls.

Raidl and Klein (43) substituted 5, 10, and 15% field pea flour in chemically leavened quick bread. The viscosity of the pea flour batters was significantly lower than either the wheat control or soy containing batters. The starch composition of the pea flour and lower water absorption properties of the protein could have affected the viscosity. Volumes of pea flour loaves were lower than the control and soy loaves. Most of the sensory characteristics of the field pea loaves were similar to those of the control quick breads. However, all flavor scores were significantly lower for pea flour products, since they had a recognizably beany or off-flavor.

Enzymatic Action in Pea Flour

Flavor is one of the major characteristics that restricts the use of legume flours and proteins in foods. Processing of soybeans, peas and other legumes often results in a wide variety of volatile compounds that contribute flavor notes, such as grassy, beany and rancid flavors. Many of the objectionable flavors come from oxidative deterioration of the unsaturated lipids. The lipoxygenase-catalyzed conversion of unsaturated fatty acids to hydroperoxides, followed by their degradation to volatile and non-volatile compounds, has been identified as one of the important sources of flavor and aroma components of fruits and vegetables. An enzyme-active system, such as raw pea flour, may have most of the necessary enzymes to produce short chain carbonyl compounds.

Lipoxygenase (linoleate:oxygen oxidoreductase) catalyzes the hydroperoxidation of fatty acids containing a methylene-interrupted conjugated diene system. The degradation of the hydroperoxides results in the formation of numerous secondary products (46-48).

The coupled oxidation of carotenoids during lipoxygenase reactions has been exploited in the baking industry for many years. Enzyme-active soy flour has been used in breadmaking since the early 1930's, when Haas and Bohn patented a process for preparing a soy flour for use in bleaching and dough improvement (49). Carotene oxidation is a secondary reaction associated with lipoxygenase (48), and the bleaching action occurs readily in a flour-water system.

Oxidative improvement of dough that contains enzyme-active flours is recognized in the bakery industry. Small quantities (less than 1% of flour weight) of enzyme-active flours result in changes in dough development profiles: higher relaxation times indicating greater dough strength, and delayed peak development providing greater tolerance to overmixing. These changes require oxygen and are related to the release of bound lipids through a lipoxygenase-coupled oxidation of the lipids. The oxidation of gluten occurs simultaneously, an effect which may also be attributable to lipoxygenase-generated products (1,50).

Unheated pea flours are also effective in bleaching and improvement of doughs (37,51). Mixing times are shorter with pea-wheat flour combinations, and mixing tolerance is increased. The levels of pea flour that are most effective for dough improvement are usually less than 5%, and 0.75 to 3% has been recommended. At higher levels, undesirable dough behavior occurs, as well as flavor deterioration.

Flavor Generation in Pea Products

The volatile constituents of raw peas have been studied with respect to the development of desirable and undesirable flavors in the processed fresh product. Numerous substances have been identified (52-54), such as ethanal, propanal, 2-trans-butenal, 2-trans-pentenal, 2-trans-hexenal, heptadienal, nonadienal, 3,5-octadecadiene-2-one, hexanal, pentanol, hexanol, pentanal, nonanal, octanal, and heptanal.

The specific compounds that are responsible for the "pea" flavor have not been identified. Bengtsson and Bosund (52) suggested that acetaldehyde, hexanal and ethanol were important, while Murray et al. (53) isolated three methoxypyrazines that have very low taste or recognition thresholds and might, therefore, be of major significance in pea flavor.

In baked products, volatile carbonyl compounds have been identified as important flavor and aroma constituents (55,56). Sosulski and Mahmoud (57) determined the composition of the major volatile carbonyls in protein supplements, fermented doughs, and in breads made from protein supplemented flours. These flours included field pea-fortified wheat flour. Some of the volatiles produced in the yeast breads are shown in Table IV. Several of the compounds associated with pea flavor are also present in the breads; their concentration is higher in soy and pea-containing breads than in unfortified wheat breads. This suggests that when unheated legume flours are used as a supplement in doughs, the resulting flavor and aroma characteristics could be a result of enzymatic activity, particularly lipoxygenase.

Table IV. Carbonyl Compounds in Yeast Bread
(mg/100 g dry weight) (57)

	Wheat	Soy[a]	Field Pea[a]
Ethanal	166	325	457
Propanal	25	56	37
2-Propanone	982	1332	1155
Butanal	25	38	37
2-Butanone	306	358	300
2-Methyl Butanal	107	840	616
Unknown	47	151	599
Hexanal	139	1096	520
Furfural + HMF	600	1470	1730

[a]Wheat flour/protein supplement/vital gluten = 83:15:2

Summary

The fortification of cereal-based products with field pea flours or protein concentrates results in an increase in both quantity and quality of protein in the food. However, the use of pea flours is limited by some of the less desirable effects. At low levels of fortification (0.75-1%), unheated pea flour is an effective dough improver, improving mixing time and tolerance, and providing bleaching action through lipoxygenase activity. At slightly higher levels, 3 to 6%, it can be used as a non-fat dry milk replacer, although this may require some additives such as vital gluten or potassium bromate. At levels above 8%, changes in crumb quality appear, and at levels of 15% and more, where the protein supplementation effect is significant, volume, flavor, aroma and overall acceptability are altered. Heating pea flour or concentrates improves flavor characteristics, but the heated product may not retain the desirable functional properties. Therefore, although field pea flours and protein concentrates have some technological and economic advantages, their potential use in food products will be limited until the functionality and flavor problems can be resolved.

Literature Cited

1. Rackis, J. J. In "Enzymes in Food and Beverage Processing"; Ory, R. L.; St. Angelo, A. J., Eds.; ACS SYMPOSIUM SERIES No. 47, American Chemical Society: Washington, D.C., 1977; pp. 244-265.
2. Pimentel, D.; Drischilo, W.; Krummel, J.; Kutzman, J. *Science* 1975, 190, 754.
3. "National Food Review," Economics, Statistics and Cooperatives, Service A. NFR-9 (Winter), 1980, p. 51.
4. Youngs, C. G. In "Oilseeds and Pulse Crops in Western Canada - A Symposium"; Western Cooperative Fertilizers, Ltd.: Calgary, Alberta, Canada, 1977; Chap. 27.

5. Reichert, R. D.; Youngs, C. G. Cereal Chem. 1978, 55, 469.
6. Tyler, R. T.; Panchuk, B. D. Cereal Chem. 1984, 61, 192.
7. Watt, B. K.; Merrill, A. "Composition of Foods," Handbook
    No. 8; U.S. Department of Agriculture, 1963.
8. Reichert, R. D.; MacKenzie, S. L. J. Agric. Food Chem. 1982,
    30, 312.
9. Vose, J. R. Cereal Chem. 1977, 54, 1141.
10. Slinkard, A. E. "Production, Utilization and Marketing of
    Field Peas"; Development Centre, University of Saskatchewan:
    Saskatoon, Saskatchewan, Canada, 1977; Ann. Report No. 1.
11. Gottschalk, W.; Mueller, H. P.; Wolff, G. Egypt. J. Genet.
    Cytol. 1975, 4, 453.
12. Bramsnaes, F; Olsen, H. S. J. Amer. Oil Chem. Soc. 1979, 56,
    450.
13. Vose, J. R. Cereal Chem. 1980, 57, 406.
14. McLean, L. A.; Sosulski, F. W.; Youngs, C. G. Can. J. Plant
    Sci. 1974, 54, 301.
15. Holl, F. B.; Vose, J. R. Can. J. Plant Sci. 1980, 60, 1109.
16. Eppendorfer, W. H.; Bille, S. W. Plant and Soil 1974, 41, 33.
17. Robertson, R. N.; Highkin, H. R.; Smydzuk, J.; Went, F. W.
    Aust. J. Biol. Sci. 1962, 15, 1.
18. Reichert, R. D. J. Food Sci. 1982, 47, 1263.
19. Holt, N. W.; Sosulski, F. W. Can. J. Plant Sci. 1979, 59, 653.
20. Evans, I. M.; Boulter, D. J. J. Sci. Food Agric. 1980, 31, 238.
21. Derbyshire, E.; Wright, D. J.; Boulter, D. Phytochemistry
    1976, 15, 3.
22. "Pea Flour and Pea Protein Concentrates," PFPS Bulletin No. 1,
    Prairie Regional Laboratory, National Research Council and
    College of Home Economics, University of Saskatchewan,
    Saskatoon, Canada, 1974; pp. 617-632.
23. Vose, J. R.; Basterrechea, M. J.; Gorin, P.A.J.; Finlayson, A.
    J.; Youngs, C. G. Cereal Chem. 1976, 53, 928.
24. Sosulski, F. W.; Elkowics, L.; Reichert, R. D. J. Food Sci.
    1982, 47, 498.
25. Biliaderis, C. G.; Grant, D. R. Can. Inst. Food Sci. Technol.
    J. 1979, 12, 131.
26. Biliaderis, C. G.; Grant, D. R.; Vose, J. R. Cereal Chem.
    1979, 56, 475.
27. Biliaderis, C. G.; Grant, D. R.; Vose, J. R. Cereal Chem.
    1981, 58, 496.
28. Reichert, R. D. Cereal Chem. 1981, 58, 266.
29. Sosulski, F. W.; Youngs, C. G. J. Amer. Oil Chem. Soc. 1979,
    56, 292.
30. Patel, K. M.; Bedford, C. L.; Youngs, C. G. Cereal Chem. 1980,
    57, 123.
31. Tyler, R. T.; Youngs, C. G.; Sosulski, F. W. Cereal Chem. 58,
    144.
32. Tyler, R. T.; Panchuk, B. D. Cereal Chem. 1982, 59, 31.
33. Sumner, A. K.; Nielsen, M. A.; Youngs, C. G. J. Food Sci.
    1981, 46, 364.
34. Johnson, D. W. J. Am. Oil Chem. Soc. 1969, 47, 402.
35. Quinn, J. R.; Paton, D. Cereal Chem. 1979, 56, 38.
36. Vaisey, M.; Tassos, L.; McDonald, B. E. Can. Inst. Food Sci.
    Technol. J. 1975, 8(2), 74.

37. Jeffers, H. C.; Rubenthaler, G. L.; Finney, P. L.; Anderson, P.
     D.; Buinsmas, B. L. Baker's Dig. 1978, 52(6), 36.
38. Hannigan, K. J. Food Engineering Int'l. 1979, 4(2), 22.
39. Nielsen, M. A.; Sumner, A. K.; Whalley, L. L. Cereal Chem.
     1980, 57, 203.
40. Tripathi, B. D.; Daté, W. B. Indian Food Packer 1975, 29(3),
     66.
41. Sosulski, F. W.; Fleming, S. E. Baker's Dig. 1979, 53(6), 20.
42. Repetsky, J. A.; Klein, B. P. J. Food Sci. 1981, 47, 326.
43. Raidl, M. A.; Klein, B. P. Cereal Chem. 1983, 60, 367.
44. McWatters, K. H. Cereal Chem. 1980, 57, 223.
45. Fleming, S. E.; Sosulski, F. W. Cereal Chem. 1977, 54, 1124.
46. Eskin, N.A.M.; Grossman, S.; Pinsky, A. CRC Crit. Rev. Food
     Sci. Nutr. 1977, 9, 1.
47. Vliegenthart, J.F.G.; Veldink, G. A. In "Free Radicals in
     Biology"; Pryor, W. A., Ed.; Academic: New York, 1982; Vol. V,
     pp. 29-64.
48. Klein, B. P.; King, D.; Grossman, S. Adv. Free Radical Biol.
     and Med. 1985, 1, 309.
49. Wolf, W. J. J. Agric. Food Chem. 1975, 23, 136.
50. Frazier, P. J. Baker's Dig. 1979, 53(12), 8.
51. American Institute of Baking, Report to Dumas Seed Company,
     1978.
52. Bengtsson, B.; Bosund, I. Food Technol. 1964, 18, 773.
53. Murray, K. E.; Shipton, J.; Whitfield, F. B.; Last, J. H. J.
     Sci. Food Agric. 1976, 27, 1093.
54. Ralls, J. W.; McFadden, W. H.; Seifert, R. M.; Black, D. R.;
     Kilpatrick, P. W. J. Food Sci. 1965, 30, 228.
55. Lorenz, K.; Maga, J. A. J. Agric. Food Chem. 1972, 20, 211.
56. Ng, H.; Reed, D. J.; Pence, J. W. Cereal Chem. 1960, 37, 638.
57. Sosulski, F. W.; Mahmoud, R. M. Cereal Chem. 1979, 56, 533.

RECEIVED December 26, 1985

4

Applications of Vegetable Food Proteins in Traditional Foods

E. W. Lusas and K. C. Rhee

Food Protein Research and Development Center, Texas A&M University, College Station, TX 77843

On a world wide basis, man obtains approximately 70% of his daily protein intake from
plant sources and 30% from animal and fish
sources. These figures are 50 and 50%, respectively, for the developed nations, and 83
and 17% for the developing countries. Oilseeds
and pulses (dry beans, lentils and peas) are
concentrated sources of proteins, and are
expected to play increasingly important roles
in human nutrition as world population grows.
    Whole oilseeds and legumes and their
derivatives (defatted flours, and protein
concentrates and isolates) are used in
traditional foods as sources of protein and
for their texture-modifying functions. This
article reviews, on a comparative basis,
processes for preparation of vegetable food
proteins, compositions and characteristics
of the resulting food ingredients, and their
functionalities and uses in traditional foods.

The pulses and certain oilseeds (soy, peanuts, sunflower seed,
sesame, and glandless cottonseed) were first accepted by man for
their storage stability, high nutrition-to-weight ratio, and attrac-
tiveness of the foods that can be made from them.
    Much of the current interest in uses of derived oilseed proteins
in compounded foods stems from projects in the mid-1960's to alle-
viate massive world hunger. Perhaps the best known of these was the
development of Incaparina at the Institute for Nutrition of Central
America and Panama, in Guatamala by Bressani and coworkers (1).
However, many other vegetable protein-enriched mass feeding foods
also have been developed, and have been reviewed (2, 3).
    In developing low cost mass feeding foods, attempts were made to
use locally available oilseed cakes and meals whenever possible. In
time, interest turned to the extraction of high protein content

ingredients from other processing residues and nontraditional crops. Prejudices once existed against pulses and certain oilseeds as foods of the poorest of the poor. However, with world-wide interest in physical fitness and dietary fiber, and concerns about possible relations between animal protein comsumption and atherosclerosis (4, 5), interest in food uses of vegetable food proteins is increasing.

Each ingredient in a compounded food is selected for a specific purpose. Even the less costly, low protein, ingredients play important roles as sources of total solids. For example, the Recommended Daily Allowance (RDA) of 65 g protein is the equivalent of 260 calories. In a 2600 calorie diet, this amount of protein can be diluted among a total caloric intake ten times greater. In addition to serving the function for which it was selected, each ingredient must not interfere undesireably with the functions of other ingredients also present.

PROCESSING

In its common use, the term "vegetable food protein" usually means a processed or derived oilseed ingredient, like defatted flour and the higher protein content concentrate and isolate forms. Almost every defatted, dehulled oilseed flour contains over 50% protein (dry weight basis). The terminology of soybean food proteins has essentially been adopted for other oilseeds: "protein concentrate" typically means a product containing over 70% protein (dry weight basis), and a "protein isolate" contains over 90% protein. For air-classified ingredients, "concentrate" refers to fractions that contain more protein than the original seed. Since full-fat or defatted flours, like those of soy, can impart undesireable flavors, the more purified food proteins ingredients like concentrates and isolates are preferred for certain applications.

Full-fat Products

A flowsheet for preparation of glandless cottonseed full-fat kernels and subsequent processing of defatted flours and concentrates and isolates is shown in Figure 1. This scheme, with specialized adaptations depending upon oilseed species, is typical for processing of all oilseeds.

Full-fat grits simply consist of whole and broken kernels that have been size-reduced by passing through cutting rolls or a hammer mill, and classified by sieving. Flakes are made by conditioning whole kernels or grits with moisture and heat to assist their plasticization, and then passing through narrowly-set smooth rolls to achieve the desired thickness. The advantage of grits and flakes is that flowable ingredients can still be had, even from high oil content seeds.

Typically, full-fat flours are made by hammer milling the seed to pass through 80-mesh or smaller size screens. However, grinding of oilseeds containing over 25% oil results in sticky flours. Thus, partially-defatted peanut and sunflower seed flours are made by first screw pressing the seed to reduce the oil to 6-18% fat content.

It is common practice to stabilize full-fat products by preheating the seed, or by extrusion as in the case of full-fat soybean flour (6). Heat treatment deactivates lipases and lipoxygenases

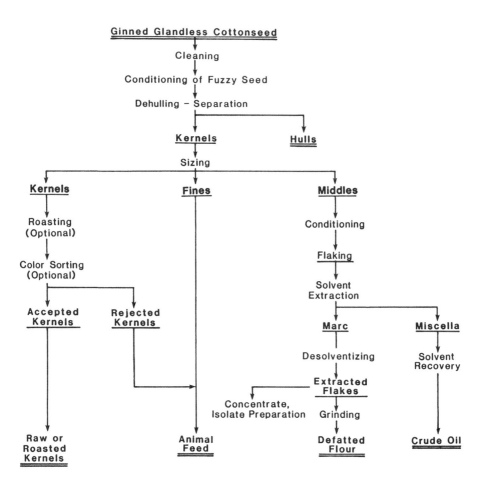

Figure 1.   General flow chart for production of glandless cottonseed
food ingredients.

which catalyze development of free fatty acids and off flavors, respectively, in addition to deactivating antinutritional factors such as trypsin inhibitors, hemagglutinins and other lectins. A limited amount of enzyme-active full-fat soybean flour is sold for bleaching and conditioning of wheat flour and bakery products (7).

High oil content seeds form pastes upon grinding, the best known example being peanut butter, which accounts for approximately 55% of domestic uses of peanuts. By law, peanut butter consists of a minimum of 90% peanuts, with the remainder being emulsifiers and/or hydrogenated fats to prevent oiling-off during storage, and salt, sweeteners and flavorings. It is typical to blanch (remove the pink/red skin or "testa"), roast, split, and remove the germ to reduce bitter flavor from peanut kernels before grinding into peanut butter.

Defatted Flours

Defatted flours are made by extracting cleaned, dehulled oilseed kernels in oilmills that are sanitary in design and operation for production of food-quality ingredients. When kernels contain less than 35% oil (like soybeans and glandless cottonseed), the seed may be conditioned, flaked and extracted directly with food grade commercial hexane. Flakes of high oil content kernels (peanuts and sunflower seed) will not remain intact during solvent extraction. It is typical to prepress these seeds to an oil content of approximately 16% and then solvent extract the broken or reflaked press cake. After countercurrent extraction, the hexane is drained and the meal desolventized and toasted by heat. The extent of toasting greatly affects protein solubility of the meal, and a range of soy flours with protein dispersibility indexes (PDI's) from 90 to 20% is available.

Defatted dehulled meals are converted into flours by ginding to pass though a 100 mesh screen. In producing sunflower flour, 95% removal of hulls (grey and white striped) from confectionery varieties, and 97% removal of hulls from oil-type (black hull) varieties is necessary to avoid noticeable grayness in the flour. Extraction also has the effect of concentrating the relative percentages of components remaining in the meal. Upon extraction, gossypol content in glandless cottonseed flour and chlorogenic acid content in sunflower seed flour can be increased by nearly 50 and 100%, respectively, from the original contents in kernels because these compounds are not soluble in hexane and stay with the meal.

Concentrates

Concentrates are made by extracting water-soluble sugars and other compounds from defatted meals or flours. This is typically a secondary extraction, using acidic ethanol-water in a chain-type or basket-type continuous extractor for processing flakes, or acidic water extraction of flour in vats, followed by spray-drying (8). Acidic polar solvents are used at or near the isoelectric point of the protein to minimize its solubility and loss. The reextracted flakes may then be ground into a flour. Concentrates are more bland than defatted flours, but still contain the fiber components of the kernel. After extraction with acidic ethanol or water, concentrates

may be neutralized to pH 6-7 to improve their solubility and functionality.

Isolates

Isolates typically are made by solublizing protein from defatted
flakes with alkali, removing the insoluble components by decanter or
desludging centrifuge, precipitating the protein at its isoelectric
pH, concentrating the precipitate by centrifugation, and spray-drying
the precipitate fraction. In some instances, pH of the acid precipitate is adjusted to near neutrality with sodium hydroxide to produce
a "proteinate". Cottonseed protein is unique in having two fractions, "storage protein" and "nonstorage protein", that can be
readily fractionated by precipitation at selected pH's (9)

Aqueous Extraction

One of the earliest methods of oil extraction practiced by man was to
mix finely ground dehulled seed in hot water and skim off the layer
of oil which separated and rose to the surface of the vat. This
process has been modernized by using mechanical grinders, stainless
steel extraction tanks, 3-phase centrifuges and spray dryers, and is
called "aqueous extraction processing" (AEP). In this procedure, the
oil is removed as an emulsion which is later broken by various means.
The protein remaining in the liquid may then be recovered and spray
dried as protein concentrates or isolates. To date, the following
oilseeds have been extracted experimentally by AEP: soybeans (10),
glandless cottonseed (11), peanuts (12), sunflower seed (13), sesame
(14), lupine (15), and coconuts (16). At the current state of the
art, minimum achievable residual oil contents in AEP concentrates
are: soybeans, 4-6%; glandless cottonseed, 6-8%; sunflower seed,
4-6%; peanuts 1-2%; and sesame 2-3%. However, the residual oils in
AEP flours, concentrates and isolates are remarkably stable.

Industrial Membrane Processing

A variety of processing options is possible through industrial
membrane processing (IMP). Ultrafiltration (UF) membranes of 20,000
molecular weight (MW) cutoff allow holding back of proteins (as
retentate), while the sugars and water-soluble compounds pass through
(as permeate). The permeate can then be processed by reverse osmosis
(RO) to obtain essentially pure water as RO permeate, and the soluble
compounds concentrated to about 20% solids as RO retentate. Experimental processes for producing soy concentrates and isolates (17),
glandless cottonseed concentrates and isolates (18), and peanut
protein concentrates and isolates (19) have been described. Various
combinations of traditional IMP and AEP/IMP techniques also have been
tried in preparation of vegetable protein concentrates and isolates.

COMPOSITION

Protein contents of selected oilseeds and legume seeds, and food
protein ingredients prepared by various procedures, are shown in
Table I. Amino acid contents and protein efficiency ratios (PER's)

Table I.   Percent Protein Content of Various Fractions of Several Oilseeds and Legumes

Fractions	Soybeans	Peanuts	Glandless Cottonseed	Sunflower Seed	Sesame	Navy Beans	Pinto Beans
Whole Seed	34	27	39	20	18	26	26
Dehulled Seed (Kernel)	43	30	43	24	26	30	29
Classical Process							
Defatted Flour	55	52	63	48	50	43	43
Protein Concentrate	72	71	71	70	77	52[c]	53[c]
Protein Isolate	93	92	91	90	90	--	--
Aqueous Extraction Process							
Protein Concentrate	68	70	67	68	71	--	--
Protein Isolate	86	92	91	89	87	--	--
Membrane Process							
Protein Concentrate	68	71	71	--	--	--	--
Protein Isolate	87	92	92[a] 80[b]	--	--	--	--

[a]Storage protein; [b]Nonstorage protein; [c]Air-classified high-protein fraction.

of selected food protein ingredients are presented in Table II.
Typical hull contents of seeds are: soybeans, 8-10%; peanuts (shells
and testas), 20-30%; fuzzy cottonseed (linters and hulls), 40-50%;
sunflower seed, 20-25%; sesame, 15-20%; and dry field beans, 8-10%.
Typical oil contents of dehulled kernels (and full-fat flours) are:
soybeans, 20-23%; peanuts, 50-55%; glandless cottonseed 35-38%;
sunflower, 50-55%; peeled sesame, 45-63%; and beans, 1-3%. Total
carbohydrate contents of defatted flours are: soybeans, 26-30%;
peanuts, 25-30%; glandless cottonseed, 23-27%; sunflower, 25-29%;
sesame, 26-30%; and beans, 60-65%. Phytate contents of defatted
flours are: soybeans 1.4-1.6%; peanuts, 1.7%; glandless cottonseed,
2.3-4.8%; sunflower, 1.5-1.9%; sesame, 3.6-5.2%; and beans, 1.4-1.8%.
Trypsin inhibitor contents of dehulled kernels are: soybeans, 4-6%;
peanuts, 0.8-1.5%; glandless cottonseed, 0.5-1.5%; sunflower, 0.7-
1.8%; sesame, 0.5-0.8%; and beans, 2-3%. The U. S. Food and Drug
Administration has set a limit of 450 ppm free gossypol in glandless
cottonseed kernels and flour, and the United Nations FAO/WHO has set
limits of 600 ppm free gossypol and 1.2% total gossypol in cottonseed
products used for human feeding.

FUNCTIONALITY

Food protein ingredients are sometimes evaluated by comparative
empirical tests, including: nitrogen solubility index (NSI) and
protein dispersibility index (PDI) profiles over a range of pH's,
water absorption, viscosity, gelling strength, whipping and foaming
capability (including volume and stability of foam); fat absorption,
and oil emulsification. Performance (including flavor, texture and
visual appeal) is often evaluated physically using standardized food
formulations, including bread (loaf volume, crumb and crust color and
texture); sugar cookies (sheet spread, surface cracking), frankfurters (fat emulsification stability, swelling and drip loss in cooking, firmness and peelability); meat loaves (moisture and fat retention during cooking); and frozen desserts (overrun and texture).
However, the most meaningful evaluations are direct in-product
trials.
        Proteins historically have been classified on the basis of their
solubility in water (albumins); salt solution (globulins); alcohol
(prolamines) and alkali (glutelins) (20).
        Texture functionality of food proteins is affected by many
factors, including relative proportions of the subfractions recovered
by extraction, and by solubility as affected by heating or toasting.
Also, it should be remembered that most food products are complex
systems with intrinsic pH and salt solubilization effects, and that
heat during product processing may coagulate and/or reduce solubility
of all proteins present, regardless of source. Solubility curves of
proteins from six raw flour sources are shown in Figure 2.

UTILIZATION

Full-Fat Products

Nut uses of roasted peanuts and sunflower kernels and deep fat fried
soybean "nuts" are well known.
        A substantial amount of vegetable protein is consumed in the

Table II. Essential Amino Acid Profiles (g/16g N) and Protein Efficiency Ratios of Various Protein Food Ingredients

	Essential Amino Acid								
	Lys	Leu	Val	Ileu	Thr	Phe+Tyr	Met+Cys	Try	PER
FAO/WHO Reference Protein	5.5	7.0	5.0	4.0	4.0	6.0	3.5	1.0	--
Soybean									
Defatted Flour	6.9	7.7	5.4	5.1	4.3	8.9	3.2	1.3	2.2
Concentrate	6.3	7.8	4.9	4.8	4.2	9.1	3.0	1.5	1.8
Isolate	6.1	7.7	4.8	4.9	3.7	9.1	2.1	1.4	1.6
Peanut									
Defatted Flour	3.0	6.4	5.3	3.2	2.6	8.4	1.9	1.0	1.8
Concentrate	3.0	6.7	4.5	4.3	2.5	10.0	2.4	1.1	1.6
Isolate	3.0	6.6	4.4	3.6	2.5	9.9	2.4	1.0	1.4
Glandless Cottonseed									
Defatted Flour	4.0	6.0	4.5	3.1	3.2	8.5	3.8	1.5	2.2
Concentrate	4.0	6.2	4.9	3.2	3.1	8.8	3.5	1.5	2.0
Isolate, Classical	4.0	6.1	4.7	3.2	3.2	8.7	3.7	1.5	1.8
Nonstorage Protein	6.2	6.4	4.6	3.4	3.4	7.4	5.0	1.6	2.4
Storage Protein	2.9	5.6	4.6	3.0	2.6	9.2	2.4	1.0	1.6
Sunflower Seed									
Defatted Flour	4.2	7.2	5.7	4.5	3.9	8.7	3.6	1.1	2.1
Concentrate	4.2	6.9	5.7	4.6	3.7	8.7	3.6	1.0	2.0
Isolate	4.1	6.4	5.5	4.3	3.7	8.6	3.4	1.0	1.9
Sesame									
Defatted Flour	3.5	7.4	4.6	4.7	3.9	10.6	5.6	1.9	1.7
Concentrate	3.0	7.1	4.5	4.2	3.6	8.4	4.9	1.9	1.6
Isolate	2.1	6.6	4.6	3.6	3.3	7.9	3.7	1.8	1.4
Whole Navy Bean Flour	7.2	7.6	4.6	4.2	4.0	7.7	1.9	1.0	--
Whole Pinto Bean Flour	6.5	7.5	5.4	5.0	3.8	6.7	1.8	0.9	--

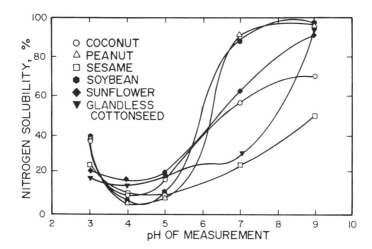

Figure 2.   Protein solubility of six defatted oilseed flours at
various pH's.

form of whole seeds and full-fat products. Examples of whole seed uses include peanuts in nut-form or in confections, sunflower seed of the loose-shelled "confectionery" type (21); sesame seed used as bread and bun toppings and for breads and buns (22); roasted soybean nuts (7); and the more recently introduced glandless cottonseed kernels used in confections, toppings for ice cream novelties and salad bars, and in specialty breads like Proteina bread.

The Food Protein Research and Development Center at Texas A&M University has developed a cookbook of glandless cottonseed kernel uses in a variety of appetizer, salad, main course, side dish, and dessert products (23).

The only full-fat oilseed flour with significant domestic sales is soy. It has been used in bakery products, breakfast cereals, canned baby foods, canned infant formulas for lactose-intolerant babies, and adult dietary beverages (24).

Bakery Products

Baked goods are the oldest known compounded foods made by mankind. Each ingredient is selected for one or more specific purposes based on contribution to functionality and compatibility, and on relative cost. Bakery products formulators are receptive to new ideas, and vegetable proteins (primarily flours and concentrates) have been well-accepted when they show a cost advantage, for example, soy flours as replacements for dried nonfat milk solids and dried eggs.

Defatted flours are especially attractive as protein sources , since 10-12% substitution of wheat flour with 50% protein flour will raise total protein content of typical wheat breads by approximately 50%, and 25% substitution will almost double the protein content of cookies. Preparation of protein-enriched breads has been reported in the literature using soy flours and protein concentrates (25), peanut flours and peanut protein concentrates (26, 27), glandless cottonseed flours, concentrates and isolates (28), sunflower seed flours and seed protein concentrates (27) and sesame flours and protein concentrates (26).

Generally, vegetable food protein ingredients are more absorbant than other dough components, with the result that mixing time and loaf volume is decreased. In addition, pan bread crumb becomes coarser and occasionally darker in color. Negative effects on loaf volume appear to be inversely related to phytic acid content.

The maximum amounts of vegetable food protein flours that can be substituted in bread without affecting loaf volume and texture are 5-10% (depending upon the source), and 18-20% can be substituted in cookies without affecting spread and surface characteristics (26). The quantity of vegetable protein flour that can be accommodated in bread can be increased substantially by pre-toasting and by the use of approximately 1.5% sodium stearoyl 2-lactylate (28) and other emulsifiers.

Breakfast Cereals

Soy flours and concentrates are used in compounded breakfast cereals, primarily for improving total protein content and PER. In the absence of dry nonfat milk solids, glucose is often included in bakery products formulations to impart a toasted brown color. Most

ready-to-eat breakfast cereals are either extruded directly from
doughs, or are first pelletized by extrusion, then flaked by rolls
before toasting in continuous ovens. Thus, it is relatively simple
to incorporate vegetable protein ingredients in these products.

Extruded Products

Soy proteins are commonly extruded as intermediate forms for later
use in processed foods. Flours and concentrates are texturized to
resemble meat chunks and are sold under the names of Texturized
Vegetable Protein (TVP), or texturized soy protein (TSP). After
rehydration with water (to approximately 18% protein and 60-65%
moisture content), up to 30% reconstituted soy protein can be used in
ground meat blends in the school lunch program, and in military and
other federal-sponsored feeding programs. These products are also
used as meat enhancers in standard of identity canned stews and
chili, and as meat extenders and replacers in nonstandardized products such as pizza toppings and sauces, and in "meatless" products
like taco filling and "Sloppy Joes". Textured peanut (29), sunflower
seed, and glandless cottonseed (30) flour products have been prepared
experimentally, demonstrating the versatility of extrusion texturiza-
tion.

Processed Meat Products

The U. S. Department of Agriculture permits up to 3.5% soy flour or
soy concentrate in standard of identity frankfurters, up to 8% soy
flour in scrapple and chili con carne, and up to 2% soy protein
isolate (containing titanium dioxide-TiO_2 as a tracer material) in
standard of identity frankfurters. Soy flours and concentrates can
bind up to 3 times their weight of water, compared to nonfat dry milk
solids which typically bind only equal weights of water. A general
practice in evaluating new vegetable food protein sources is to
compare their performance to soy flour in frankfurters (31). An
extruded soy protein isolate fiber product is also used for structur-
ing mechanically deboned meats, poultry, fish and seafoods into
rolls, sticks or fillets, or into extruded shrimp shapes.
     USDA regulations also allow use of non-meat proteins in products
such as pumped ham and corned beef, provided the finished product
contains a minimum protein content of 17%. Pumping to achieve a
cooked yield of 130% is permitted (32).

Dairy Products

Cow's milk, extended with full-fat soy flour, is produced by CIATECH
in Chihuahua, Mexico, and a peanut isolate-extended water buffalo
milk ("Miltone") has been produced in India for approximately 20
years (33). Establishment of soymilk plants in Southeast Asia and
Latin America is a growth industry and soy milks also are sold in the
United States. Various beverages, flavored to mask the taste of
soybeans, have been introduced world-wide during the last two decades. A major problem of vegetable proteins is that, being globulins, they are readily precipitated by calcium fortification, required under domestic law for milk replacement products. Whereas consumers in other countries readily accept shaking of beverage containers

before drinking, the domestic market prefers products which do not settle. Research progress has been made on succinylation and maleylation of soy, peanut and glandless cottonseed proteins to prevent their precipitation in the presence of calcium fortification (34).

Considerable interest has been shown in uses of vegetable food proteins in cheese-type products. Attempts have been made to coprecipitate casein and vegetable protein in the typical vat process for making cheeses (35). Rhee (36) has found that up to 50% peanut protein isolate and 25% soybean isolate can be effectively substituted for sodium caseinate in the preparation of imitation cheeses.

Summary

The invention of new food forms is not required to increase uses of vegetable food proteins in the American diet. Uses of flours, contrates and isolates continue to grow as increasingly more convenience foods are formulated and produced in factories, either for grocery or institutional sales.

REFERENCES

1. Bressani, R.; Elias, L. G.; Aguirre, A.; Scrimshaw, N. S., J. Nutr. 1961, 74, 201-208.
2. Orr, E., "The Use of Protein-rich Foods for the Relief of Malnutrition in Developing Countries: An Analysis of Experience"; Tropical Products Institute Monograph G 73, 1972; Aug.
3. Aguilera, J. M.; Lusas, E. W., J. Am. Oil Chem. Soc. 1981, 58(3), 514-520.
4. Carroll, K. K., J. Am. Oil Chem. Soc. 1981, 58(3), 416-419.
5. Kritchevsky, D., J. Am. Oil Chem. Soc. 1979, 56(3), 135-140.
6. Mustakas, G. C.; Albrecht, W. J.; Bookwalter, G. N.; McGee, J. E.; Kwolek, W. F.; Griffin, E. L., Jr. Food Technol. 1970, 24, 1290-1296.
7. Circle, S. J.; Smith, A. K., In "Soybeans: Chemistry and Technology"; Smith, A. K.; Circle, S. J., Eds; AVI Publ. Co., Westport, 1978.
8. Campbell, M. F.; Kraut, C. W.; Yackel, W. C.; Yang, H. S. In "New Protein Foods"; Altschul, A. M.; Wilcke, H. L., Eds.; Academic Press, New York, 1985; Chap. IX.
9. Martinez, W. H.; Hopkins, D. T. In "14th Nutritional Quality of Foods and Feed. Part II. Quality Factors: Plant Breeding, Composition, Processing, and Anti-Nutrients."; Friedman, M., Ed.; Marcel Dekker, New York, 1975; pp. 355-374.
10. Rhee, K. C.; Cater, C. M.; Mattil, K. F. U. S. Patent 4 151 310, 1979.
11. Cater, C. M.; Rhee, K. C.; Hagenmaier, R. D.; Mattil, K. F., J. Am. Oil Chem. Soc. 1974, 51(4), 137-141.
12. Rhee, K. C.; Mattil, K. F.; Cater, C. M., Food Eng. 1973, 45(5), 82-86.

13. Hagenmaier, R. D., J. Am. Oil Chem. Soc. 1974, 51(10), 470-471.
14. Chen, S. L. M.S. Thesis, Texas A&M University, College Station, 1976.
15. Aquilera, J. M.; Gerngross, M. F.; Lusas, E. W., J. Fd. Technol. 1983, 18, 327-333.
16. Rhee, K. C.; Lusas, E. W. In "Tropical Foods: Chemistry and Nutrition"; Inglett, G. E.; Charalambous, G., Eds.; Academic Press, New York, 1979, Vol.2, pp. 463-484.
17. Lawhon, J. T.; Lusas, E. W., Food Technol. 1984, 38(12), 97-106.
18. Lawhon, J. T.; Manak, L. J.; Lusas, E. W., J. Food Sci. 1980, 45, 197-203.
19. Lawhon, J. T.; Manak, L. J.; Rhee, K. C.; Lusas, E. W., J. Food Sci. 1981, 46, 391-398.
20. Vickery, H. B., Physiol. Revs. 1945, 25, 347.
21. Lusas, E. W. In "New Protein Foods"; Altschul, A. M.; Wilcke, H. L. Eds.; Academic Press, New York, 1985; Chap. XII.
22. Johnson, L. A.; Sulerman, T. M.; Lusas, E. W., J. Am. Oil Chem. Soc. 1979, 56(3), 463-468.
23. Simmons, R. G.; Golightly, N. H., "Cottonseed Cookery"; Food Protein R&D Center, Texas A&M University, College Station, 1981.
24. Dubois, D. K.; Hoover, W. J., J. Am. Oil Chem. Soc. 1981, 58(3), 343-346.
25. Rooney, L. W.; Gustafson, C. B.; Clark, S. P.; Cater, C. M., J. Food Sci. 1972, 37, 14-18.
26. Khan, M. N.; Rhee, K. C.; Rooney, L. W.; Cater, C. M., J. Food Sci. 1975, 40(2), 580-583.
27. Khan, M. N.; Lawhon, J. T.; Rooney, L. W.; Cater, C. M., Cereal Chem. 1976, 53(3), 388-396.
28. Khan, M. N.; Wan, P. J.; Rooney, L. W.; Lusas, E. W., Cereal Foods World 1980, 25(7), 402-404.
29. Aquilera, J. M.; Rossi, F.; Hiche, E.; Chichester, C. O., J. Food Sci. 1980, 45(2), 246-254.
30. Taranto, M. V.; Meinke, W. W.; Cater, C. M.; Mattil, K. F., J. Food Sci. 1975, 40, 1264-1269.
31. Smith, G. C.; Juhn, H. I.; Carpenter, Z. L.; Mattil, K. F.; Cater, C. M., J. Food Sci. 1973, 38, 849-855.
32. "National School Lunch Program: Special Food Service Program for Children"; Federal Register 39(60), 1197.
33. Chandrasekhara, M. R.; Ramanna, B. R.; Jagannath, K. S.; Ramanathan, P. R., Food Technol. 1971, 25, 596-598.
34. Choi, K. R.; Lusas, E. W.; Rhee, K. C., J. Food Sci. 1982, 47, 1713-1716.
35. Rhee, K. C.; Lusas, E. W., In "Annual Report"; Food Protein R&D Center, Texas A&M University, College Station, 1985.
36. Rhee, K. C., In "Annual Report"; Food Protein R&D Center, Texas A&M University, College Station, 1985.

RECEIVED January 24, 1986

Uses of Soybeans as Foods in the West with Emphasis on Tofu and Tempeh

Hwa L. Wang

Northern Regional Research Center, Agricultural Research Service, U.S. Department of Agriculture, Peoria, IL 61604

Soybeans have been used as food in the Orient since ancient times and various methods have been developed to make soybeans as palatable as possible. In recent years, a large number of these simply processed soyfoods are emerging in the West. Tofu and tempeh are the most popular and have the fastest growth rate of any soyfood in America. Tofu is made by coagulating the protein with a calcium or magnesium salt from a hot-water extracted, protein-oil emulsion of whole soybeans. It is a highly hydrated gelatinous product with a bland taste. The texture characteristics of the curds vary from soft to firm, depending on the processing conditions. Thus, tofu can be easily incorporated with other foodstuffs and used in nearly every culinary context from salad to dessert and from breakfast foods to dinner entrees. Tempeh is made by fermenting cooked soybeans with a mold, Rhizopus oligosporus. The white mycelium covers the bean mass and binds it into a firm cake that can be sliced, seasoned, and cooked just like meat. Tempeh is becoming a hamburger alternative for vegetarians.

In the West soybeans have been primarily viewed as an oilseed. As early as 1908, some European countries started to import beans from China to process into oil and meal. Commercial oil mill processing plants, however, were not built in the U.S. until 1922 (1). The oil was then mostly for industrial uses. But, because of declining industrial uses and increasing demand for edible oil in the late 1930s, research on soybean oil for food uses was encouraged. By the seventies, soybean oil became a major edible oil in the United States. Soybean meal, a by-product from the extraction of oil has been widely used as animal feed since the late 1930s. American soybean processors also produce a variety of edible protein products from the meal, such as defatted grits and flours, concentrates, and isolates. These products became known in the fifties and reached the highest popularity as meat extenders in 1973. Since then their use has been static, although the food industry continues to use these products as ingredients in many food systems. The use of these edible soy protein products as direct food, however, is still waiting to be accepted.

In East Asia, on the other hand, soybeans have traditionally
been used directly as foods. Centuries of creative striving have
yielded great numbers of protein foods that are versatile, easily
digestible and delicious. It has been said that because of the
existence of soybeans, the countries of East Asia succeeded in
supporting a high population density in those distant days.

Based on processing technology, the soybean foods that have
been consumed in East Asia may be classified into two general types:
non-fermented and fermented (2-8) as shown in Table I. Names of
these foods and the details of preparing and serving such foods may
vary from country to country. Among them, soybean curd (tofu) and
soy sauce have been the most widely consumed in the Orient.

Table I. Oriental Soybean Foods

Foods	Description and Uses
Nonfermented	
Fresh green soybeans	Picked plump, firm, bright green before maturation. Cooked and served as fresh green vegetable.
Soybean sprouts	Bright yellow beans with 3-5 cm sprouts. Cooked and served as vegetable or in salad.
Soybean milk	Water extract of soybeans, resembling dairy milk. Served as breakfast drink.
Protein-lipid film	Cream-yellow film formed over the surface of simmering soybean milk. Cooked and used as meat.
Soybean curd (tofu)	White or pale yellow curd cubes coagulated from soybean milk. Served as main dish with or without further cooking.
Soybean flour	Ground roasted dry beans, nutty flavor. Used as filling or coating for pastries.
Fermented	
Soy sauce	Dark reddish brown liquid, salty taste suggesting the quality of meat extract, a flavoring agent.
Miso	Paste, smooth or chunky, light yellow to dark reddish brown, salty and strongly flavored resembling soy sauce, a flavoring agent.
Hamanatto	Nearly black soft beans, salty flavor resembling soy sauce, a condiment.
Sufu	Cream cheese-type cubes, salty, a condiment, served with or without further cooking.
Tempeh	Cooked soft beans bound together by mycelium as a cake, clean fresh and yeasty odor. Cooked and served as main dish or snack.
Natto	Cooked beans bound together by and covered with viscous, sticky polymers produced by bacteria, ammonium odor, musty flavor, served with or without further cooking as main dish or snack.

According to industry statistics gathered in early 1984 by Shurtleff and Aoyagi of the Soyfood Center in California (9), Americans were consuming an average of 2.22 pounds of such traditional soybean foods per year per capita as compared to 1.37 pounds of the modern soy protein foods. The annual production of tofu has increased from 12,020 MT in 1978 to 24,300 MT in 1983, with an average annual growth rate of 15% and the highest growth rate of 27% in 1979. The annual production of tempeh has increased from 511 MT in 1981 to 900 MT in 1983, with the average annual growth rate of 33% and the highest growth rate of 36% in 1982. Soy protein isolates, which had the fastest growth rate among the modern soy products, increased from 11,000 MT produced in 1970 to 41,000 MT in 1982, with an average annual growth rate of 11%. The production figures on soy isolates also include exports. Consequently, the growth rate of consumption in U.S. would be significantly lower than the growth rate of production indicated. Furthermore, there has been little or no growth in the combined U.S. production of soy flour, isolates and concentrates since 1974 based on a survey made by Shurtleff and Aoyagi (9).

Tofu

Tofu has long been a source of protein in the Orient. It has much the same importance to the people of the Orient that meats, eggs and cheese have for the people in Western Countries. Tofu is usually sold in the form of a wet cake with a creamy-white color, smooth custard-like texture and a bland taste. It is highly hydrated and, depending on the water content, tofu products with different characteristics can be produced. The typical oriental type of tofu has a water content about 85%. Japanese prefer tofu having a smooth, fragile texture that contains about 88% water. The Chinese, on the other hand, produce many types of firm products with a chewy meat-like texture and a water content as low as 50-60%. Western consumers like tofu with a firm texture; therefore, tofu found in the U.S. supermarkets contains 75-80% water.

Because of its fine texture, bland taste and light color, tofu has been used in nearly every culinary context: desserts, salads, breakfast foods, dinner entrees and burgers. It can be cooked simply with desired flavoring agents or it can be easily incorporated with other foodstuffs.

Preparation. Tofu is made by coagulating the proteins with a calcium or magnesium salt from a hot-water extracted, protein-oil emulsion (soybean milk) of soybeans. The process is simple (Figure 1), but making a reproducible high-quality product is a problem. Many factors, from the quality of the dry beans to pressing the curd can affect the yield and quality of the resultant tofu. In recent years, several studies (10-14) have been made on tofu processing in an attempt to better understand the process and to optimize the processing conditions.

Three main steps are involved in making tofu (Figure 1): Preparation of soybean milk, coagulation of protein, and formation of tofu cakes in a mold. By experience, the Orientals have found that the most suitable ratio of water (including that absorbed during soaking) to dry soybeans is 8:1 to 10:1. Watanabe et al (15)

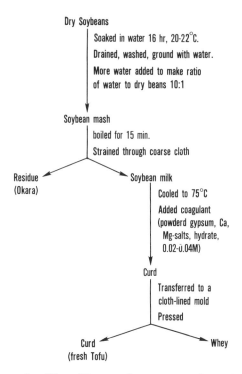

Figure 1.   Flow diagram for preparation of tofu.

noted a significant reduction in the amount of protein and total
solids extracted when the amount of water used was reduced to 6.5
times that of dry beans. Increasing the amount of water over 10:1
increases the extractable materials; however, excess water would
result in a soybean milk too low in protein to achieve a proper curd
formation.

Soaking the beans in water facilitates the grinding and it
removes some undesirable factors such as the gas-forming oligosaccharides, but it also leaches out soluble proteins. To keep
soaking losses at a minimum and to save energy, hydration of soybeans
at an ambient temperature, around 20-22°C for 16-18 hr, is most
suitable (16). Grinding the soaked beans expedites the extraction
and also the formation of the protein-lipid emulsion. A heat
treatment is essential, not only for protein denaturation to attain
proper curd formation (17), but to improve nutritional value and to
reduce off-flavor. Based on in vitro digestibility and amino acid
composition (13), the maximum nutritive value of soybean milk can be
ensured by boiling for 10-15 min. Excessive heat not only adversely
affects the nutritive value and tofu texture, but also reduces the
total solids recovery, and thus reduces the tofu yield.

Coagulation is the most important step in terms of reproducible
yield and texture of tofu, but it is the least understood. In the
Orient, making of tofu has been considered an art, and even today,
the relationship between the ion binding to the soybean proteins and
the coagulation phenomenon are still not completely understood.
According to Fukushima (17), native soy protein molecules are
unfolded during heating. Consequently, the free SH groups, disulfide
bonds and hydrophobic groups are exposed. In a dilute solution, the
unfolded proteins remain soluble, but as the exposed groups are
brought closer together through concentration by drying or freezing,
or through neutralization of molecular charges, irreversible
aggregates result. The bonds responsible for the intermolecular
polymerization are the disulfide bonds formed by the sulfhydryl/
disulfide interchange reaction and also the interactions among the
hydrophobic amino acid residues. Fukushima (17) postulated that the
irreversible coagulation in tofu production is brought about by
decreasing molecular charges, because added Ca^{++} or Mg^{++} ions bind
with the negatively charged acidic amino acid residues and the
sulfide group of the unfolded protein molecules.

Recent studies (10,11,13) have shown that both ionic
concentration and type of coagulant affect the quantity and quality
of the resultant tofu. Results obtained from our laboratory are
shown in Figures 2 and 3 (13). When the concentration of the
coagulant is lower than 0.01 M and higher than 0.1 M, there is no
curd formation. In studying the binding of unfractionated soybean
proteins with calcium ion, Appurao and Rao (18) observed that at
higher calcium concentrations the extent of precipitation decreases
and the protein becomes soluble again. Our data are consistent with
their observations. Data in Figure 2 also show that salt
concentrations between 0.02-0.04 M result in the highest nitrogen
recovery and that the sensitivity to the concentration shifts is the
least. Thus, the use of salt at a level between 0.02 to 0.04 M is
most likely to yield a reproducible product with a high nitrogen

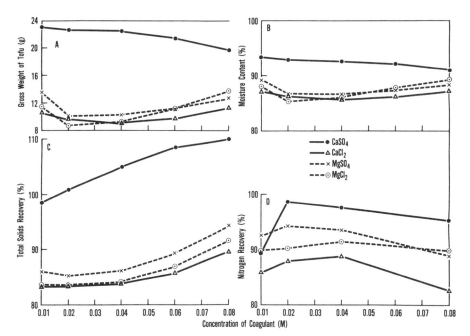

Figure 2.   Relationship of concentration and type of coagulant
to the yield of tofu (13).

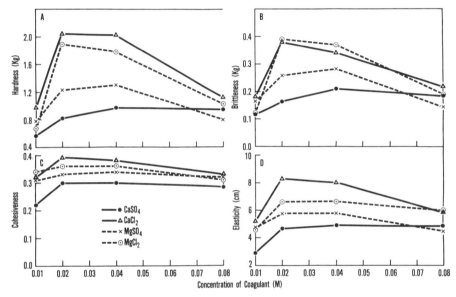

Figure 3.   Relationship of concentration and type of coagulant
to the texture characteristics of tofu (13).

content. Data on texture characteristics (Figure 3) also indicate
that there is little sensitivity to concentration changes when a
salt level of 0.02-0.04 M is used. The hardness and the brittleness
of the curds, however, are influenced by the type of salt used.
Calcium chloride and magnesium chloride result in curds with much
greater hardness and brittleness than calcium sulfate and magnesium
sulfate suggesting that anions have a greater effect on texture than
cations. This observation agrees with a study by Aoki (19) on the
effect of salt on the gelation of soybean proteins, where anions
were found to have a stronger effect on water-holding capacity than
cations. The hardness of tofu increases as its water content
decreases (14). Tsai et al (11) found that coagulant concentrations
between 0.025 and 0.03 M are the most suitable for making Chinese-
style tofu.

The temperature of the soybean milk at the time coagulants are
added, the modes of mixing and pressing greatly affect the yield
and texture of the resulting tofu. Increasing the temperature
increases the hardness, but decreases the volume, weight, and water
content of tofu (13). Increasing mixing also decreases tofu volume
and increases hardness (10,13).

Thus, many factors affect the final product. By knowing the
effects that each factor produces, one can choose and establish a
set of conditions to reproduce the desired type of tofu.

Soybean Variety. Saio and her coworkers (20) speculated that soybean
variety could have an effect on tofu texture, because they found
that a gel made from isolated 11S globulin is much harder and more
elastic than that made from 7S globulin. They also noted increasing
tofu hardness as the amount of phytic acid was increased in soybean
milk, but such chemical variations between varieties may not be
great enough to have the influence on texture that processing
variables do. Recently, Skurray et al (12) made a study with 15
varieties and found no significant correlation between the ratio of
7S to 11S protein or phosphorus content and the quality of tofu.
However, they did find that the quality of tofu is more affected by
the amount of calcium ion added. Wang et al (14) studied varietal
effects with 5 U.S. and 5 Japanese soybean varieties grown under the
same environmental conditions, and found that the composition and
color of tofu are affected by soybean variety but that yield and
texture are not significantly affected. Varieties with a dark brown
hilum result in tofu with a less attractive color so that these
varieties are not desirable. Tofu made from varieties with a high
protein content has a higher protein/oil ratio than tofu made from
varieties with less protein (Table II). Therefore, varieties with
high protein are preferred.

Table II.   Protein to Oil Ratio of Tofu and Soymilk
as Affected by Protein and Oil Content of Soybeans (14)

Variety	Soybeans[a] Protein %	Oil %	Protein/Oil Soybeans	Soymilk	Tofu
Wase-Kogane	45.2	17.4	2.60	2.49	2.07
Vinton	45.1	17.9	2.52	2.50	2.01
Toyosuzu	44.1	18.1	2.44	2.13	1.87
Coles	43.2	18.5	2.34	2.11	1.78
Yuuzuru	42.3	17.7	2.39	2.30	1.89
Tokachi-Nagaha	41.8	17.3	2.42	2.12	1.88
Weber	40.9	19.3	2.12	1.75	1.57
Hodgson	40.9	19.4	2.11	1.90	1.67
Corsoy	40.8	18.9	2.16	1.95	1.69
Kitamusume	40.8	19.4	2.10	1.86	1.57

[a]Dry basis.

Composition and Nutritional Value of Tofu. The composition of tofu
may vary depending on soybean variety used and method of preparation
as exemplified in Tables II and III. Since the method of preparation
greatly affects the water content of the product, it influences the
percentage of other components (Table III). Tofu available in U.S.
supermarkets usually contains 80% or less water so that it may have
more than 10% of protein. Other nutrients typically present in 100
g tofu with 84.8% of water are: fiber, 0.1 g; calcium, 128 mg;
phosphorus, 126 mg; iron, 1.9 mg; sodium 7 mg; potassium, 42 mg;
thiamin, 0.06 mg; riboflavin, 0.03 mg; niacin, 0.1 mg (21). Tofu
has been a source of calcium in the Oriental diet. The calcium
content of tofu varies depending on the coagulant used. Tseng et al
(22) reported that tofu prepared with a calcium salt has a higher
calcium content and higher Ca/P ratio than that prepared by other
coagulants. They suggested that tofu can help to correct the
imbalanced Ca/P ratio in many American diets. Also, tofu made from
calcium salt is a good source of calcium in vegetarian diets.

Table III. Composition of Tofu as Related to Percentage of Water

Water %	Protein %	Oil %	Other Solids %	Protein Oil	Ref.
84.8	7.8	4.2	3.2	1.9	21
85.1	7.5	4.2	3.2	1.8	14[a]
(84.2-85.7)	(6.8-8.4)	(3.8-4.7)		(1.6-2.1)	
88.0	6.0	3.5	2.5	1.7	17

[a]Average of 10 soybean varieties, with ranges in parenthesis.

Although tofu has been claimed as a low-calorie protein food,
the following comparison needs to be considered. One hundred grams
of tofu (water, 84.8 g; protein 7.8 g; oil, 4.2 g) contains about 72
calories, whereas 100 g of cooked hamburger (water, 54.2 g; protein,
24.2 g; fat, 20.3 g) has 286 calories (21). Although the hamburger

meat provides more than three times as much protein as tofu, it has
a lower protein/fat ratio (1.2 vs 1.9). Accordingly, hamburger has
more calories than tofu based on the weight that provides the same
amount of protein. Per 50 g protein, hamburger has 591 and tofu has
461 calories. Hamburger also has more fat in such a comparison.
However, protein/fat ratio varies greatly among the various cuts and
types of meat. Meat that is well trimmed to arrive at a higher
protein/fat ratio could have less calories and fat content than tofu
to provide the same amount of protein. Therefore, tofu can not
always be considered as low calorie food. However, tofu is a low-
density protein food, and is thus more filling. The indisputable
nutritional assets of tofu are the absence of cholesterol and
lactose, and low amounts of saturated fatty acids.

Microbiological Quality of Tofu. Tofu is a protein-rich substrate
with pH around 6, hence it is quite susceptible to microbial growth.
Traditionally, tofu has been made and consumed in the same day.
However, in the United States, tofu may be held at the supermarkets
for many days on produce counters before consumption where
temperatures are usually 10-15°C. Thus, microbial deterioration
becomes a serious problem (23-25).
    Tofu should be relatively free of vegetative microbial cells if
it is made under proper sanitary conditions. Cooking the soybean
mash at the boiling temperature for 15 min should kill all vegetative
cells and leave only the heat-resistant spores as survivors. However,
the presence of heat-resistant, spore-forming bacteria observed on
soybeans (unpublished data) suggests that, even though contamination
may have been prevented during processing, bacterial growth could
occur if tofu is stored under conditions suitable for the microbes
to grow. Measures then must be taken to prevent the growth of these
microorganisms in order to improve the microbiological quality of
tofu. In addition to proper storage conditions, the processors
should throughly clean the beans to reduce the surface microbial
load and carry out the processing with a high level of sanitary
practices (25,26).
    Recently, studies to evaluate the microbiological safety of
tofu were made by Kovats et al (27). Water-packed tofu samples were
inoculated with such common food pathogens as Clostridium botulinum,
Staphylococcus aureus, Salmonella typhimurium, and Yersinia
enterocolitica, then held at different temperatures for various
lengths of time. They found that all four organisms grew in water-
packed tofu. C. botulinum toxin was produced in tofu held at 15°
and 25°C within 3 days and 1 wk, respectively, but not at 5° and
10°C within 6 wk. S. aureus and S. typhimurium grew at similar
rates at 10, 15, 25°C, but neither pathogen grew during storage at
5°C. Staphylococcal enterotoxin was not produced within 4 wk at
10°C even though a population of greater than 10^7/g was present in
most samples analyzed. Y. enterocolitica grew at all temperatures
evaluated (5, 10, 15 and 25°C). Isolates recovered from tofu samples
agglutinated with antiserum (WA-SAA), indicating that the isolates
continued to express their virulence-associated determinant after
growing in tofu. Thus, like many other foods, the potential of
microbial hazards is great for tofu produced under unsanitary
conditions and/or stored at improper temperatures. High level

sanitary practices, pasteurization after packaging, and storage and display at 5°C or less by manufacturers, distributors and retailers were recommended by Kovats et al (27). Tofu can be kept frozen or freeze-dried to prevent microbial deterioration. However intermolecular interactions occur during frozen storage. As a result, the texture of tofu is changed from soft, smooth to sponge-like with a meat-like chewiness.

Tempeh
Tempeh, originating in Indonesia, is made by fermenting dehulled and briefly cooked soybeans with Rhizopus mold; the mycelium binds the soybean cotyledons together in a firm cake. Freshly fermented tempeh has a clean, yeasty odor. When sliced and deep-fat fried, it has a nutty flavor and pleasant aroma. Tempeh is used as a main dish and meat substitute in Indonesia. Vegetarians in the West have used tempeh as hamburger patties. Unlike most other fermented soybean foods which usually involve more than one microorganism, long brining, and an aging process, tempeh fermentation is short and simple and requires only one mold.

Preparation. Traditionally, soaked, hand-dehulled and briefly boiled soybeans are inoculated with small pieces of tempeh from a previous fermentation, wrapped in banana leaves which also serve as a source of inoculum, then left at room temperature for 1-2 days. Studies carried out by Hesseltine et al (28) resulted in a pure culture fermentation as shown in Figure 4. To save time and labor, mechanically dehulled, full-fat grits have replaced the traditional whole soybeans. A tempeh starter containing spores of Rhizopus oligosporus NRRL 2710 (29) is now used in the West in place of traditional inocula. Not only are Petri dishes the most convenient laboratory container, they also are used commercially in preparing tempeh patties. Other containers such as shallow aluminum foil or metal trays with perforated bottoms and perforated plastic film covers, and perforated plastic bags and tubings have been used successfully for tempeh fermentation. Rhizopus oligosporus requires air to grow, but too much aeration will cause spore formation and also may dry up the beans, resulting in poor mold growth. Therefore, both properly perforating the containers and packing the beans for fermentation are important.

Tempeh Products from Grains and Other Beans. Traditionally tempeh is made from soybeans known as tempeh kedele. However, copra (pressed coconut cake) and the by-product from making soybean milk have also been used in Indonesia to make tempeh known as tempeh bongkrek and tempeh gembus, respectively. Recently, attempts have been made (30) to make tempeh-like products from grains such as wheat, oats, barley, rice, mixtures of cereal and soybeans, and from beans other than soybeans, such as broad beans, cowpeas, mung beans and winged beans. In the United States, tempeh made from a mixture of wheat and soybeans (31) has been available commercially since 1970.

Biochemical Changes During Fermentation. The effects of R. oligosporus on soybeans have been studied by several investigators (30,32). As shown in Table IV, the fermentation process does not

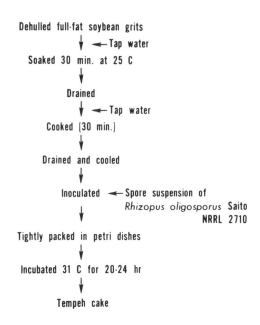

Figure 4.   Flow diagram for tempeh fermentation.

greatly affect the proximate composition of soybeans. The slight
increase in the percentage of protein reflects the decrease of other
constituents that the mold might have consumed for growth.

Table IV.   Proximate Composition of Soybeans and Tempeh

Food	Protein %	Oil %	Fiber %	Ash %	Carbohydrates %
Soybeans[a]	47.8	26.8	3.9	3.4	18.1
Tempeh	48.1	24.7	3.1	3.3	20.8

[a]Treated similarly as for fermentation except the inoculation step
was omitted.

As the mold begins to grow rapidly, the temperature of fermenting
beans rises a few degrees above the incubator temperature, then
falls as the growth of mold subsides. The pH increases steadily to
above 7, presumably because of protein break-down. After 69 hrs. of
incubation, soluble solids rise from 13 to 28%, soluble nitrogen
increases from 0.5 to 2.0%, but total nitrogen remains fairly constant
and reducing substances decrease slightly, probably due to utilization
by the mold. The mold does not utilize the carbohydrates in the
soybeans; instead, it uses the soybean oil as its energy source.

Although total nitrogen remains fairly constant during
fermentation, free amino acids increase in tempeh. The essential
amino acid index, on the other hand, is not significantly changed by
fermentation. Perhaps the amount of mycelial protein present in
tempeh is not high enough to alter greatly the amino acid composition
of the soybeans, nor does the mold depend upon any specific amino
acid for growth.

Niacin, riboflavin, pantothenic acid and vitamin B_6 contents
are greatly increased in tempeh during fermentation, whereas thiamin
exhibits no significant change. R. oligosporus appears to have a
great synthetic capacity for niacin, riboflavin, pantothenic acid,
and vitamin B_6, but not for thiamin.

The most interesting and important finding was the presence of
vitamin B_{12} in tempeh because foods derived from plant materials are
deficient in this essential nutrient. Vitamin B_{12} is known to be
synthesized by microorganisms; however, molds have not been reported
to produce Vitamin B_{12}. Liem et al (33) found a fairly high amount
of vitamin B_{12} in commercial tempeh bought from Canada and
subsequently confirmed that the major source of the vitamin was a
result of a contaminating bacterium which the authors isolated and
identified as Klebsiella. They reported that tempeh made from pure
mold isolated from commercial tempeh contained nutritionally
insignificant amounts of vitamin B_{12}, confirming that the tempeh
mold does not produce the vitamin. On the other hand, tempeh made
with the mold and the bacterium, Klebsiella, isolated from commercial
tempeh, had 150 ng of vitamin B_{12} per gram of tempeh. The presence
of the mold does not interfere with the production of vitamin B_{12} by
the bacteria, but presence of the bacteria requires longer
fermentation time. Liem and his co-workers also demonstrated that
soaking soybeans either with or without an acid did not increase the
vitamin B_{12} content. The results indicated that tempeh made with

pure mold fermentation under hygienic conditions adopted for food
processing in this country has no nutritionally significant amount
of vitamin B_{12}. However, there is a great potential to make vitamin
B_{12}-enriched tempeh with an inoculum containing R. oligosporus and a
vitamin B_{12}-producing bacterium.

Nutritional Value. Although enzymes produced by the mold have acted
upon the substrate and partly hydrolyzed its constituents into small
molecules, the digestibility coefficient of tempeh tested by the
rat-assay method is not significantly different from that of
unfermented substrate. It is also not surprising to learn that the
protein efficiency ratio (PER) of tempeh (8) as determined by rat
assay is not significantly different from that of unfermented but
properly heat-treated soybeans, because the fermentation process
does not signficantly change the total nitrogen and the amino acid
composition. However, tempeh made from a mixture of wheat and
soybeans has been shown (8) to have a better protein value than that
made from soybeans alone, because of the complementary effect of
mixed proteins and the increased availability of lysine in wheat
from fermentation.

Increase in vitamins, such as niacin, riboflavin, pantothenic
acid, Vitamin B_6, and Vitamin B_{12}, is of great nutritional
significance, especially where fortifying foods with synthetic
vitamins is not practiced.

Microbiological Quality of Tempeh. Like tofu, tempeh should normally
be relatively free of contaminated vegetative cells, but may not be
free of heat-resistant spores. Failure of fermentation caused by
bacterial contamination has been reported by tempeh producers. In
order to assure successful fermentation, a starter with high viability
is as important as a high level of sanitary practices. To maintain
the microbiological quality of tempeh, steaming after fermentation
and then freezing are recommended. Results obtained from studies on
the safety of tempeh inoculated with different bacterial pathogens
(34) indicated that tempeh should be steamed after fermentation and
then kept at 5°C or below until it is used.

Conclusions

Whole soybean foods have been the major source of protein in East
Asia since ancient times and various methods have been developed to
make soybeans more palatable. Among these simply made soybean
foods, tofu and tempeh have recently become increasingly popular in
the West. The production processes may not improve the nutritional
value of soybean protein, but they reduce the cooking time, improve
the organoleptic characteristics and increase the versatility of
soybean uses. With the recently increasing interest in protein
foods other than those from animal origin, the consumption of tofu
and tempeh has been on an upsurge in the West and is expected to
continue its steady growth in the years to come.

Literature Cited

1. Smith, A. K.; Circle, S. J. Eds. In "Soybeans: Chemistry and
    Technology"; AVI: Westport, Conn., 1972; Chap. 1.
2. Hesseltine, C. W.; Wang, H. L. In "Soybeans: Chemistry and
    Technology"; Smith, A. K.; Circle, S. J., Eds.; AVI: Westport,
    Conn., 1972; Chap 11.
3. Wang, H. L.; Mustakas, G. C.; Wolf, W. J.; Wang, L. C.;
    Hesseltine, C. W.; Bagley, E. B. "Soybeans as Human Food:
    Unprocessed and Simply Processed," U.S. Department of
    Agriculture, Utilization Res. Rep. No. 5, January, 1979.
4. Fukushima, D. J. Am. Oil Chem. Soc. 1979, 56(3), 357.
5. Winarno, F. G. J. Am. Oil Chem. Soc. 1979, 56(3), 363.
6. Wang, H. L.; Hesseltine, C. W. In "Microbial Technology';
    Peppler, H. J.; Perlman, D., Eds.; Academic Press: New York,
    1979; Vol II, Chap. 4.
7. Fukushima, D. "Proc. World Conference on Soya Processing and
    Utilization," J. Am. Oil Chem. Soc. 1981, 58, 346.
8. Wang, H. L. In "Handbook of Processing and Utilization of
    Agriculture," Wolff, I. A., Ed.; CRC Press, Boca Raton, Florida,
    1983; Vol. II, Part 2, p. 91.
9. Shurtleff, W.; Aoyagi, A. "The Soyfoods Industry and Market:
    Director and Databook." The Soyfoods Center, Lafayette, CA.
    1984-1985.
10. Saio, K. Cereal Foods World 1979, 24, 342.
11. Tsai, S. J.; Lau, C. T.; Kao, C. S.; Chen, S. C. J. Food Sci.
    1981, 46, 1734.
12. Skurray, G.; Cunich, J.; Carter, O. Food Chem. 1980, 6, 89.
13. Wang, H. L.; Hesseltine, C. W. Process Biochem. 1982, 17, 7.
14. Wang, H. L.; Swain, E. W.; Kwolek, W. F.; Fehr, W. R.
    Cereal Chem. 1983, 60, 185.
15. Watanabe, T.; Fukamachi, C.; Nakayama, O.; Teramachi, Y.; Abe,
    K.; Suruga, S.; Mivanage, S. The Report of Food Research
    Institute, Ministry of Agriculture and Forestry, Japan, 1960,
    1413 (in Japanese).
16. Wang, H. L.; Swain, E. W.; Hesseltine, C. W.; Heath, H. D.
    J. Food Sci. 1979, 44, 1510.
17. Fukushima, D. In "Chemical Deterioration of Proteins"; Whitaker,
    J. R.; Fugimaki, M., Eds.; ACS SYMPOSIUM SERIES No. 123, American
    Chemical Society: Washington, D.C., 1980, p. 211.
18. Appurao, A. G.; Rao, M. S. Cereal Chem. 1975, 52, 21.
19. Aoki, H. Nippon Hogli Kagaku Kaishi 1965, 39, 277.
20. Saio, K.; Kamiya, M.; Watanabe, T. Agri. Biol. Chem. 1969, 33,
    1301.
21. "Composition of Foods," Agriculture Handbook No. 8, U.S.
    Department of Agriculture, 1975.
22. Tseng, R. Y. L.; Smith-Nury, E.; Chang, Y. S. Home Economics
    Res. J. 1977, 6, 171.
23. Hankin, L.; Hanna, J. G. Bulletin 810, The Connecticut
    Agriculture Experimentation Station, New Haven, Connecticut.
    1983, March.
24. Aulisio, C. C. G.; Stanfield, J. T.; Weagant, S. D.; Hill, W.
    E. J. Food Protect. 1983, 3, 226.
25. Rehberger, T. G.; Wilson, L. A.; Glatz, B. A. J. Food Protect.
    1984, 3, 177.

26. Wang, H. L. J. Am. Oil Chem. Soc. 1983, 61, 528.
27. Kovats, S. K.; Doyle, M. P.; Tanaka, N. J. Food Protect. 1984,
    47, 618.
28. Hesseltine, C. W.; Smith, M.; Bradle, B.; Ko Swan Djien.
    Dev. Ind. Microbiol. 1963, 4, 275.
29. Wang, H. L.; Swain, E. W.; Hesseltine, C. W. J. Food Sci.
    1975, 40, 168.
30. Steinkraus, K. H., Ed. Handbook of Indigenous Fermented Foods,
    Marcel Dekker, New York, 1983.
31. Hesseltine, C. W.; Smith, B.; Wang, H. L. Development in
    Industrial Microbiology 1967, 8, 179.
32. Wang, H. L.; Hesseltine, C. W. In "Microbial Technology";
    Peppler, H. J.; Perlman, D., Eds.; Academic Press: New York,
    1979; Vol. II, p. 95.
33. Liem, I. T. H.; Steinkraus, K. H.; Cronk, T. C. Appl. Environ.
    Microbiol. 1977, 34, 773.
34. Tanaka, N.; Kovats, S. K.; Guggisberg, J. A.; Meske, L. M.;
    Doyle, M. P. J. Food Protect. 1985, 48, 438.

RECEIVED December 13, 1985

Incorporation of Cottonseed into Foods for Humans

Elwood F. Reber

Department of Nutrition and Food Sciences, Texas Woman's University, Denton, TX 76204

The Food and Drug Administration approved the use of cottonseed containing not more than 450 ppm gossypol for human use. Glandless whole kernel cottonseed flakes and cot-n-nuts are commercially available. Studies incorporating cottonseed into many different foods have yielded acceptable products with improved protein quantity and quality. The presence of free gossypol and cyclopropenoid fatty acids (CPFA) potentially limits the use of cottonseed in human foods. The levels of free gossypol and CPFA are reduced in processing the seed and preparation of food. The amount of free gossypol and CPFA in the food as eaten should be determined. The American Oil Chemists' Society method for free and total gossypol is not specific for gossypol and gave false positive readings for several food ingredients. Glandless cottonseed is a valuable addition to the food supply in the United States. The development and utilization of glandless cottonseed in the rest of the world would be a major contribution to the alleviation of severe hunger in some areas.

Glanded cottonseed is the cotton crop grown around the world. The cotton plant and the cottonseed have pigment glands which contain several pigments that can make the seed appear dark green to black. One of the pigments is gossypol. Cottonseed was first suggested as a food source for human consumption in 1876 (1). When glanded cottonseed is fed to monogastric animals, the oxygen-carrying capacity of the blood is reduced and shortness of breath, edema of the lungs and paralysis may occur. Rats usually show signs of loss of appetite, decreased growth, rough hair coat and listlessness (2). However, the toxic effects of gossypol affects various animals to different degrees of severity. The Chinese (3) found the consumption of unrefined cotton oil by humans caused a reversible infertility in males. Cottonseed flour has been included in INCAPARINA baby food formulas in Guatemala (4).

0097–6156/86/0312–0061$06.00/0
© 1986 American Chemical Society

Glanded cottonseed has been used to produce a defatted cottonseed flour with reduced gossypol content by a procedure known as the
liquid cyclone process (LCP). LCP cottonseed flour has been used in
the preparation of many foods that have been tested in several animal
and human nutrition studies. The commercial production of LCP cottonseed flour has not been successful (5).

Studies with Raw, Cooked and Roasted Cottonseed

Glandless cottonseed is obtained from a cotton variety that has
reduced amounts of pigment and gossypol in the seed kernel. The
first report of a sub-acute toxicity investigation of glandless cottonseed fed to rats was made by Reber and Pyke (6). The investigation was designed to satisfy the requirements of the Food and Drug
Administration (FDA). The results were submitted to the FDA, who
objected to the validity of the study on the basis that cottonseed
oil was a derived component of cottonseed kernels; so the control
diet did not serve its purpose. The FDA suggested that another study
be conducted where the rats would be fed a control diet containing 6%
corn oil to be able to conclude without reservation that glandless
cottonseed kernels are safe for human consumption, based on rat
studies.
A protocol approved by the FDA to determine the safety of low
gossypol cottonseed kernels for human consumption was the basis for
the second study by Reber (7). To prepare raw cottonseed flour, raw
kernels were ground to meet Ro-tap sieve specifications of lab chow.
To prepare roasted cottonseed flour, raw kernels were dry roasted at
not less than 121°C for not less than 5 min. To prepare cooked cottonseed flour, raw kernels were cooked in steam until batch temperature had been at or above 121°C for 5 min. All cottonseed kernels
were ground in the manner described above. The kernels contained not
more than 0.037% (370 ppm) of free gossypol. They were free of Salmonella and did not contain detectable amounts of aflatoxin. The
proximate analyses of the cottonseed flours are shown in Table I.
The glandless cottonseed was obtained from Rogers Cottonseed Co.,
Waco, Texas, then processed and analyzed by the Food Protein Research
and Development Center, Texas A&M University, College Station, Texas.

Table I.   Analysis of Cottonseed Flours[a]

	Raw	Roasted	Cooked
Moisture, %	6.20	2.27	4.23
Protein, %	39.13	40.56	39.81
Oil, %	35.45	37.34	36.94
Ash, %	4.30	4.44	4.39
Crude Fiber, %	1.48	1.42	1.37
Gossypol (free) %	0.037	0.03	0.034
Gossypol (total) %	0.042		0.034
Lead, ppm	1.5	1.5	1.5
Arsenic, ppm	0.1	0.1	0.1
Heavy metals, ppm	10.0	10.0	10.0

[a]Reprinted from J. Food Sci. 1981. 46(2):593-596.   Copyright by
Institute of Food Technologists.

Growing female rats utilized cooked or roasted cottonseed more efficiently than raw cottonseed or control diet. The percentages of pups alive at birth surviving to 4 days were significantly higher for rats fed raw or cooked cottonseed than roasted cottonseed. There were no significant differences due to diet observed in average body weights of dams at parturition and at weaning time or in weight of offspring.

There were no detrimental effects due to feeding low gossypol (370 mg/kg) cottonseed kernels at a level of 20% of the diet equivalent to 74 mg of free gossypol per kg of diet as eaten. Rats fed cottonseed grew as well as or better than control animals. Heat treatment of cottonseed apparently made one or more nutrients more available on a nutritional basis to female rats. Overall the diet containing cooked cottonseed appeared to be a better diet than the diet containing roasted cottonseed. These observations led to an investigation of the protein quality of the cottonseed as affected by the processing.

The same shipment of raw, cooked and roasted glandless whole kernel cottonseed flours used in the FDA study was used to determine the protein efficiency ratio (PER) of each flour (8). The adjusted PER (Table II) of cooked (2.10) cottonseed was significantly higher than roasted (1.77) cottonseed. Protein retention efficiency (PRE) for roasted cottonseed (58.08) was lower than values for raw (60.54) and cooked (62.95) cottonseed. Relative protein values (RPV) indicated a utilization of 91, 91 and 96% of the protein in raw, roasted and cooked cottonseed, respectively. The multiplication of the (RPV) percentage utilization and the protein content of the cottonseed (Table I) results in the relative utilizable protein values (Table II).

Table II. Average Protein Efficiency Ratio (PER), Relative Protein Value (RPV) and Relative Utilizable Protein for Raw, Roasted and Cooked Whole Kernel Cottonseed[a]

Parameter	Casein	Cottonseed		
		Raw	Roasted	Cooked
PER (adjusted)	2.50[b]	1.93[c]	1.77[ce]	2.10[cd]
RPV	----	0.91	0.91	0.96
Relative Utilizable Protein[f]	----	35.61	36.90	38.21

[a]Reprinted from J. Food Quality 1983. 6:65-71.

[b]Was significantly (P<0.01) higher than c for each parameter (Duncan's multiple analysis).

[d]Was significantly (P<0.01) higher than e for each parameter (Duncan's multiple analysis).

[f]RPV X protein content.

Cooked glandless cottonseed protein quality was superior to that of raw and roasted cottonseed. The roasting process adversely affected protein quality. Supplementation of roasted cottonseed with 0, 0.2, 0.4, 0.6 and 0.8% L-lysine indicated a peak PER response at 0.45%. The adverse effect of roasting on the protein quality of cottonseed was overcome by addition of L-lysine which made the protein

quality of roasted cottonseed comparable to that of cooked cottonseed (Table III). The PER study indicated that glandless cottonseed
kernels are a feasible protein source. The protein in cottonseed is
of greater value and utilized to a greater extent after the kernel
is cooked.

Table III. Average Adjusted Protein Efficiency Ratio for Raw, Cooked
and Roasted Cottonseed Supplemented with Lysine[a]

	PER
Casein	2.50[b]
Raw Cottonseed	1.89[c]
Cooked Cottonseed	1.87[c]
Roasted Cottonseed	1.69[c]
Roasted +0.2% lysine	1.85[c]
Roasted +0.4% lysine	1.95[c]
Roasted +0.6% lysine	1.92[c]
Roasted +0.8% lysine	1.83[c]

[a]Reprinted from J. Food Quality 1983. 6:65-71.

[b]Significantly (P<0.01) higher than [c].

Cottonseed in Nigerian Foods

High-protein, low-cost Nigerian foods, chin-chin, puff-puff, akara
(9), akamu, sugar cookies and yeast bread (10) were developed using
raw full-fat cottonseed (Table I), defatted cottonseed (Table V),
soybean, peanut or sesame flours as protein supplements.
The foods were evaluated by African and non-African panels for
appearance, texture, absence of greasiness, palatability and overall
acceptability. The African panel rated all the food products significantly (P=0.01) higher for all characteristics than the non-African
panel. The high scores may be attributed to the foods being cul-
turally and traditionally acceptable. Akara, puff-puff and chin-chin
are deep fat fried Nigerian foods. Akara is the most common food
product in Africa containing cowpeas. Deep fat fried balls prepared
from akara provide a tasty, fresh product available at specific times
to the consumer. The size of akara balls determines whether they are
to be eaten as an appetizer or as a main dish. The nutritional value
of cowpeas is associated with a high protein value, ranging from 20-
25%. The akara products in which soybean or sesame flour replaced
25% of the cowpeas were rated more desirable more frequently by both
panels than were the original cowpea akara or the akara in which
either cottonseed flour or peanut flour was used.
Chin-chin is crisp, slightly sweet, golden brown knots of pas-
try. These deep fat fried cakes are made in different shapes and
styles. Chin-chin is often flavored with vanilla extract, nutmeg,
caraway seed, orange or lemon rind. The African panel preferred the
chin-chin with no substitution for all characteristics except appearance. The non-African panel preferred chin-chin without any substitutions for all characteristics. The soybean flour was the most
desirable substitution at 30% of the all purpose flour.
Puff-puff, soft, golden brown balls of wheat flour batter fried
in deep fat, is a major snack food in most African countries. The

African panel preferred the puff-puff containing 30% defatted cottonseed. The non-African panel preferred the puff-puff containing the
all purpose flour for all characteristics except the absence of
greasiness.

The substitution of the seed flours for cowpeas or wheat flour
increased the percentage protein (Table IV) in all food products and
increased the chemical scores of the limiting amino acids,
methionine and cystine, for all foods.

The percentage of protein and fat in the total solids of food
products was calculated on the basis of the recipes prior to deep fat
frying (Table IV). Following the frying of the products proximate
analyses were obtained and the protein and fat in the total solids
were calculated. There was a large increase in the fat content of
all products as a result of the deep fat frying. In all fried food
products the percentage of fat exceeded the percentage of protein
(Table IV). The addition of the seed flours to akara did not significantly increase protein quality of the product containing cowpea
seed protein. Although akara was high in fat content after frying
the high level of protein in akara resulted in PER levels above 2.00.
In both the chin-chin and puff-puff products the fat to protein ratio
was considerably higher than for akara (Table IV). The high fat content in the diet reduced the food intake and the weight gained by the
rats. Thus, the PER values were much lower than would be expected.
Puff-puff was not a good source of protein and supplementation with
seed flours made it an even poorer source due to the high fat content
in the product as eaten. The addition of seed flours to chin-chin
improved the protein quality with defatted cottonseed producing the
best improvement. Supplementation of chin-chin, therefore, would be
advantageous for the Nigerian people.

Table IV. Percentage Protein and Fat of Total Solids of the Food
Before and After Frying[a]

Products[b]	Before		After	
	Protein	Fat	Protein	Fat
Akara				
100% cowpeas	27.5	6.0	20.9	24.9
75% cowpeas/25% FC	30.8	14.0	21.7	36.9
75% cowpeas/25% DC	34.7	6.8	24.0	32.0
Chin-chin				
100% wheat flour	11.1	21.0	9.8	32.0
70% wheat/30% FC	16.7	27.5	13.6	40.4
70% wheat/30% DC	20.8	21.3	16.4	30.3
Puff-puff				
100% wheat flour	17.3	7.5	14.0	30.6
70% wheat/30% FC	22.4	13.7	17.0	41.9
70% wheat/30% DC	25.4	7.9	18.7	30.6

[a]Reprinted from J. Food Sci. 1983. 48(1):217-219. Copyright by
Institute of Food Technologists.

[b]FC, full-fat cottonseed; DC, defatted cottonseed.

Akamu, which is made from fermented corn, millet or sorghum is an important traditional breakfast and weaning cereal in most African countries. The African panel preferred the 100% cornstarch akamu, whereas, the non-African panel preferred the 25% full fat cottonseed akamu for texture, palatability and overall acceptability. In general, the cornstarch akamu and 25% soybean akamu products were rated higher than the 25% sesame akamu for certain characteristics. The 25% defatted cottonseed akamu and 25% sesame seed akamu were rated lower than the akamu containing full fat cottonseed, soybeans, peanuts or cornstarch flours. Replacing wheat flour with various flours in yeast bread produced an acceptable bread, but with decreased volume. Bread containing 15% full fat cottonseed was rated the highest for moistness. However, both panels rated the 15% defatted cottonseed and 15% sesame seed bread lower compared to the yeast bread made from 15% full fat cottonseed, soybean, peanut or wheat flours. When sugar cookies were evaluated, the 30% defatted cottonseed cookies were rated lower by both panels compared to the cookies made from other flours. Overall, the African panel rated the akamu, yeast bread and sugar cookies higher than the non-African panel for all characteristics evaluated. The defatted cottonseed products were rated lower for all three products compared to the other flour products.

Cottonseed, peanut, sesame seed, and soybean flours, when used as supplements, add to the quality of the protein. The improvement of protein quality was due to the flours compensating for limiting amino acids. Supplementation using various flours improved total protein, amino acid content and some physical characteristics in akamu, yeast bread and sugar cookies.

Cottonseed in American Foods

Food grade cottonseed flakes and cot-n-nut kernels are available commercially. The fuzzy cottonseed with a moisture content of 9 to 10% and a maximum free gossypol content of 300 ppm is used to prepare the cot-n-nuts and flakes. The seed is cleaned and processed through a cottonseed huller. The dehulled seeds are sized, gravity graded and aspirated. Cot-n-nuts are prepared by placing the seeds in a dry tumbling roaster set at 154°C for 5 min. After the seeds are roasted, they are electronically color-sorted to produce an end product with a maximum free gossypol content of 100 ppm. The cot-n-nuts are vacuum packed in 50 lb. bags. Raw cottonseed which contains 100 ppm or less free gossypol is used to prepare cottnseed flakes. The raw cottonseed is roasted at 121°C for 5 min. The hot cottonseed is put through a flaker and then vacuum packed in 50 lb. bags. The cottonseed flakes or cot-n-nuts may be made into a flour. Proximate analyses of the cottonseed flakes and cot-n-nuts are shown in Table V.

Cottonseed flakes and cot-n-nuts have been used in many foods. The Department of Nutrition and Food Sciences has published (11) "Cooking with Cottonseed", a booklet which is available on request. Examples of foods with cottonseed flakes are orange cotton cake, cot-n-nut chewies, granola bars, sausage and cotton balls, bolla chips (a high protein snack food), cot-n-nut cups, sausage in a cottonseed roll, and cotton cheese crackers.

Table V.   Proximate Analysis (%) of Defatted Cottonseed Flour,
Cottonseed Flakes and Cot-n-nuts

Component	Cottonseed Flour[a]	Cottonseed Flakes	Cot-n-nut
Moisture	8.20	6.00	4.70
Protein	55.94	36.06	39.06
Fat	2.16	31.20	35.40
Fiber	2.80	4.50	1.90
Ash	5.99	5.64	4.16
Nitrogen-free extract	24.91	16.60	14.78

[a]Defatted flour.

The use of cottonseed protein in human foods was described by
Bressani (12) in 1965. There are many cottonseed derivatives and
their potential food uses have been described by Berardi and Cherry
(13). These derivatives could become the cornerstone for a whole new
phase of food fortification (14). Shanklin, Hume and Gould (15)
investigated the possibility of incorporating glandless cottonseed
into products for use in school lunch programs. The products
included: pizza roll, sausage roll, heavenly angel biscuits, corn-
bread, cotton 'lasses, and cotton crispies. Using a 7-point hedonic
scale, the elementary (Grades 1-6) school children rated all products
a 5 (OK), or better for overall acceptability, with pizza rolls and
cotton crispies receiving the highest ratings. When the protein content was determined, glandless cottonseed increased the amount of
protein compared to the standard in all products except cotton
crispies. Therefore, it was considered feasible to incorporate
glandless cottonseed into these food products and obtain an acceptable product with improved protein content.

Limiting Factors

Gossypol. A factor limiting the use of cottonseed in foods for
humans is free gossypol. Cottonseed and okra belong to the Malvaceae
family. Gossypol is present in okra in small amounts (16). Okra is
consumed in considerable amounts by people in some regions but the
intake of free gossypol in the diet containing okra is unknown. The
maximum amount allowed by the FDA is 450 mg of free gossypol per kg
of cottonseed. Rogers Cottonseed Co. sets a limit of 300 mg of free
gossypol per kg of raw cottonseed and sorts the seed to produce
flakes or cot-n-nuts containing 100 mg or less of gossypol per kg of
finished product. The goal should be and is to reduce the free gossypol content of cottonseed for human consumption to zero. Until
that goal is achieved the free gossypol values in cottonseed before
incorporation into food should be used in applying the FDA safety
level. However, processing reduces free gossypol content.
    Bressani et al. (17) have shown that processing affects free
gossypol. In their study at the Institute of Nutrition of Central
America and Panama (INCAP), Vegetable Mixture 9 (28% ground corn, 28%
ground sorghum, 38% cottonseed flour, 3% torula yeast, and 3%
hydrated leaf meal) was cooked in water at 85 to 87°C for varying
times and with different additives. Free gossypol was decreased by

cooking in water and by wetting without cooking. The effects were more evident when calcium or iron salts were added or when sugars were present. The reduction of free gossypol was more pronounced when both $FeSO_4$ and $Ca(OH)_2$ were present. This suggests that an alkaline pH is important in reducing free gossypol, especially in the presence of iron. In the Bressani et al. study, the temperature used for cooking was not very high and cooking in water was the only technique used.

Reber, Hopkins and Liu (18) described the effects of food preparation procedures and of various food ingredients on the free and total gossypol contents of food products containing cottonseed. Free and total gossypol levels were determined by the method in the American Oil Chemists' Society Official Methods manual (19). Gossypol-acetic acid, purchased from the Sigma Chemical Company, St. Louis, Mo., was used for standard calibration. Free gossypol is that portion of the gossypol which is soluble in aqueous acetone. In the analytical procedure bound gossypol is released when subjected to mild acid hydrolysis. The analytical procedure is used to measure free gossypol and total gossypol. Bound gossypol is total gossypol minus free gossypol. Analysis of the flour was done by Pope Testing Laboratories, Inc., Dallas, Texas. The flour contained: 9.00% moisture, 55.41% protein, 3.24% fat, 0.80% fiber, 7.43% ash and 0.044% (440 mg/kg) free gossypol. There was 460 mg/kg free gossypol in the cottonseed flour as determined by the TWU lab. Recovery of free gossypol added to the cottonseed was 89%. Foods prepared using the cottonseed flour were orange balls, cinnamon wafers, brownies, hamburger patties, and Sloppy Joe filling. These foods were chosen because various preparation techniques including no cooking, baking, and simmering were required to make them. All of the products had been previously prepared at Texas Woman's University.

Orange balls are a no-bake cookie, therefore the gossypol analyses were conducted on the finished products. There was a loss of 4% of the free gossypol (Table VI). Brownies, cinnamon wafers, and hamburger patties were mixed, then cooked. These three foods were analyzed for gossypol content before and after cooking to determine if the amount of free gossypol changed during the preparation of the product. Hand mixing the ingredients in the brownies resulted in an increase of 26% of the free gossypol (Table VI). Mixing and cooking of the other foods caused a decrease in free gossypol. No free gossypol was found in baked hamburger patties.

The Sloppy Joe filling was analyzed only after cooking, because it was cooked as it was mixed. No free gossypol was recovered in the cooked final product. These data indicate free gossypol is reduced as a result of mixing cottonseed with other recipe ingredients, but in the foods tested, some free gossypol remained. The application of heat in cooking the foods also caused a reduction of the presence of free gossypol. In the case of hamburger patties and Sloppy Joe filling, no free gossypol was recovered. The analysis for free gossypol of foods as eaten should be the basis for measuring gossypol intake by humans.

The following food products prepared for human consumption were analyzed for free and total gossypol: Proteina bread, a product of Mrs. Baird's Bakery, Fort Worth, Texas; crunchy and smooth Cot-N-Nut Butter, processed at the American Nut Corporation, Lewisville, Texas;

Table VI. Percentage of Free Gossypol Change
in Food Products

Method of Preparation	Food Product	Change in Free Gossypol,%
Hand Mixing	Orange Balls	– 4
Hand Mixing	Cinnamon Wafers	– 13
Baking	Cinnamon Wafers	– 85
Hand Mixing	Brownies	+ 26
Baking	Brownies	– 76
Hand Mixing	Hamburger Patties	– 64
Baking	Hamburger Patties	–100
Simmering	Sloppy Joe Filling	–100

and Bolla, a fried chip prepared at Texas Woman's University. The
concentration of free gossypol in the cottonseed included in the food
was 450 mg or less per kg of cottonseed. The free gossypol in these
foods ranged from 10 to 38 mg per kg of food (Table VII).

Table VII. Products Containing Cottonseed

Product	Gossypol, mg/kg Free	Total
Proteina Bread	10	90
Crunchy Cot-N-Nut Butter	20	150
Smooth Cot-N-Nut Butter	38	180
Bolla	20	130

The total gossypol data for many foods was much higher than that
contributed by the cottonseed. The increase in free gossypol as a
result of mixing the ingredients in preparing the brownie mix was a
problem. The analytical work indicated the AOCS procedure for free
and total gossypol analysis was not specific for gossypol. The lack
of specificity had been noted in a study in which animal tissues were
analyzed for gossypol content (20). The gossypol values were attributed to some constituent other than gossypol reacting with the
reagent. The AOCS method involves a color reaction that takes place
between the aldehyde groups of gossypol and the amine group of
aniline to form dianilino-gossypol (19). Aniline is a non-specific
reactant with aromatic aldehydes. Stipanovic et al. (16) reported
false-positive free gossypol results in the analysis of okra and
glandless cottonseed. The false-positive gossypol readings appeared
to be due to degradation products of hydroxylated unsaturated fatty
acid triglycerides reacting with aniline. The free gossypol values
reported in Table VI and VII may represent free gossypol contributed
by the cottonseed flour, false-positive readings caused by the ingredients, or some of each. Therefore, it is impossible to state the
free or total gossypol content of prepared foods based on the AOCS
method.

The lack of specificity of the AOCS method produced false-
positive readings for orange juice, vanilla wafers, pecans, brown
sugar, chili powder, vanilla, cocoa, Worcestershire sauce, and black
pepper (Table VIII). Cooked ground beef, shortening, and baking
powder did not produce false-positive gossypol readings. The
increase in the number of false-positive values for total gossypol,
compared to free gossypol, may have been due to the liberation of
reactive groups as a result of the mild acid hydrolysis required in
the analysis for total gossypol.

Table VIII.   Free and Total Gossypol in Recipe Ingredients

Ingredient	Gossypol, mg/kg	
	Free	Total
Cottonseed flour	460	660
Chili powder	310	1090
Vanilla	270	550
Cocoa	125	410
Orange juice	86	90
Worcestershire sauce	80	1620
Black pepper	70	140
Brown sugar	35	390
Vanilla wafers	33	180
Pecans	8	30
Kitchen bouquet	0	1940
Granulated sugar	0	1370
Salt	0	1360
Powdered sugar	0	600
Tomato paste	0	550
Cinnamon	0	470
Egg	0	410
All purpose flour	0	150
Beef, ground, raw	0	50
Whole wheat flour	0	50
Onion	0	20

The rate of condensation of an amine ($R-NH_2$) with a carbonyl
compound (R-CHO) as a function of pH forms a bell-shaped curve over
the pH range 1 to 7 with a peak at pH 3.5. The color developed
(optical density) by the reaction at pH intervals from 3 to 11 were
determined. A bell-shaped curve did not occur. Peak optical
densities (O.D.) occurred at pH 4.00 (0.755), 5.00 (0.776) and 5.63
(0.775). Pure vanilla over a pH range of 3 to 11 had an inverted
bell-shaped curve, but had some optical density in the range of 4 to
5.65 pH. Gossypol acetate was added to the vanilla so that the gossypol concentration was the same as in the standard gossypol solution. The curve of the gossypol plus vanilla solution was similar
to the vanilla curve. Thus, vanilla may inhibit the gossypol-aniline
color development reaction in some way.
    Solutions obtained for the analysis of other food ingredients
had the greatest color development, i.e., false gossypol values, at
pH 5 to 6. False gossypol reactions could not be eliminated by
altering the pH of the reacting solutions in the AOCS (19) procedure.
It is essential to be able to accurately determine the amount of an

ingredient in a food, especially when that ingredient may have an adverse effect on an individual who consumes it. The inability to accurately analyze a human food product for free and/or total gossypol using AOCS (19) stresses the need for modification of that procedure or to establish an analytical method that is reliable.

Nomeir and Abou-Donia (21) have reported on gossypol:high performance liquid chromatographic (HPLC) analysis and stability in various solvents. Stipanovic et al. (16) suggested the analysis of gossypol be made on the basis of the amount of dianilinogossypol formed. Dianilinogossypol was measured by thin layer chromatography (TLC).

Cyclopropenoid Fatty Acids. Cottonseed contains cyclopropenoid fatty acids (CPFA) which must be investigated to determine their effects on humans and other monogastric animals. Related to this is the hepatocarcinogenicity of whole kernel glandless cottonseed and cottonseed oil in rainbow trout, Salmo gairdeneri, (22-24). However, the Food and Drug Administration (FDA) has adopted the position that fish are not sufficiently related to man to necessitate the FDA changing its acceptance of cottonseed and its byproducts in human foods (25).

Cottonseed contains two cyclopropenoid fatty acids, malvalic and sterulic acid. The presence of these two CPFA's could be a limiting factor in the use of cottonseed for human food (26). The preparation of foods containing cottonseed may decrease the amount of cyclopropenoid fatty acids.

Hampden, Reber and Stewart (27) compared the effects of feeding whole kernel cottonseed flour at 50% of the diet as the sole source of lipid and protein to feeding cottonseed oil or corn oil as the lipid source and casein as the protein source. Growing male rats utilized the diet containing cottonseed oil more efficiently (P=0.001) than male rats fed whole kernel cottonseed or corn oil diets. The whole kernel cottonseed and cottonseed oil diets had a hypocholesterolemic effect on the rats as compared to rats fed corn oil. Total serum cholesterol levels were 73.4, 63.1 and 61.8 mg/dl for rats fed corn oil, cottonseed oil and whole kernel cottonseed, respectively (28). Rats fed corn oil and whole kernel cottonseed diets had significantly (P=0.05) higher total serum triglyceride levels than rats fed cottonseed oil. Very low density and low density lipoprotein concentrations were significantly (P=0.05) higher in rats fed cottonseed oil and whole kernel cottonseed compared to rats fed corn oil. Fat depots of rats fed cottonseed oil and whole kernel cottonseed had significantly (P=0.001) higher linoleic acid levels and significantly (P=0.001) lower oleic acid levels than rats fed corn oil. No detrimental effects due to feeding whole kernel cottonseed flour at 50% of the diet were observed. There were no CPFA in fat depots of the rats. Cottonseed oil in the diet resulted in superior performance as compared to rats fed corn oil.

Heavy Metals. The levels of lead, arsenic and heavy metals for cottonseed flour, flakes and cot-n-nuts are in Table IX. Under the category of heavy metals, lead, arsenic and mercury are included. Generally, adults consume about 0.2 to 0.3 mg of lead per day; however, consumption of 2 to 3 mg per day (1 mg in children) for an

extended time can result in toxicity. To prevent accumulation of
lead in the body, the suggested maximum amount of total lead intake
for adults and children is 600 and 300 ug per day, respectively (29).
Arsenic is found in many foods at levels below 1 mg per kg (PPM).
Daily intake of arsenic from food varies from tens of micrograms to
over 1 mg (30). According to the levels in Table IX, consumption of
large amounts of cottonseed over a prolonged period of time could
result in lead, arsenic and heavy metal toxicity.

Table IX. Heavy Metal Content of Cottonseed Flakes, Cot-N-Nut
and Defatted Cottonseed Flour

Component	Cottonseed Flakes	Cot-N-Nut	Defatted Cottonseed Flour
Lead, ppm	1.5	0.9	1.5
Arsenic, ppm	< 0.5	< 0.5	< 0.5
Heavy metal, ppm	< 10	< 10	< 10

Allergens. A limiting factor in the consumption of cottonseed is the
presence of allergens. Allergens may produce immediate or delayed
symptoms. The allergen apparently is not present in the fat portion
of the cottonseed, but is in the protein which is reduced by heat,
though not completely destroyed (31). With the increased production
of products containing cottonseed, the presence of allergens becomes
important even though it may affect only a small portion of the population. In developing countries in which cereals are the major food
source of protein, allergies to cottonseed could become critical to
the production and utilization of cottonseed products, especially if
cottonseed is to become an important food source by its incorporation
into other foods.

Acknowledgments

The Natural Fibers and Food Protein Commission of Texas, Carl Cox,
executive director, supported this work over an extended period of
time, which is gratefully acknowledged. Additional support came from
Texas Cottonseed Crushers Assoc. and Cotton, Inc. The author
acknowledges and expresses appreciation for the cooperation and contributions of graduate students. The contributions of Theresa Smith
and Carol Owsiany in preparing this manuscript are appreciated.

Literature Cited

1. Spadaro, J. J.; Gardner, H. K. J. Amer. Oil Chem. Soc. 1979,
    56, 422-4.
2. Gallup, W. D. J. Biol. Chem. 1931, 91, 387-94.
3. Anonymous. China Med. J. 1979, 8, 455-8.
4. Bressani, R. "Laboratory Evaluation of Protein Rich Mixtures";
    Nutrition (Proc. 9th Intern. Cong. Nutr., Mexico, 1972): S.
    Karger: Basel, Switzerland, 1975; Vol. IV.
5. Anonymous. "The Economic Feasibility of Modifying and Operating
    the Lubbock Cottonseed Flour Plant to Supply Potential Domestic
    Markets"; Experience, Inc.: Minneapolis, Minnesota, 1978.
6. Reber, E. F.; Pyke, R. E. J. Food Safety 1980, 2, 87-95.

7. Reber, E. F. J. Food Sci. 1981, 46, 593-6.
8. Reber, E. F.; Kuo, M. F.; Sadeghian, S. B. J. Food Quality
    1983, 6, 65-71.
9. Reber, E. F.; Eboh, L.; Aladeselu, A; Brown, W. A.; Marshall,
    D. C. J. Food Sci. 1983, 48, 217-19.
10. Eboh, L. Ph.D. Thesis, Texas Woman's University, Texas, 1980.
11. Shanklin, C. W.; Gould, R. "Cooking with Cottonseed"; Texas
    Woman's University: Denton, Texas, 1984.
12. Bressani, R. Food Tech. 1965, 19, 1655-61.
13. Berardi, L. C.; Cherry, J. P. Cotton Gin and Oil Press 1979.
14. Editorial. Food Engineering 1979.
15. Shanklin, C. W.; Hume, F. K.; Gould, R. School Food Service
    Res. Rev. 1984, 8, 119-21.
16. Stipanovic, R. D.; Donovan, J. C.; Bell, A. A.; Martin, F. W.
    J. Agric. Food Chem. 1984, 32, 809-10.
17. Bressani, R.; Elias, L. G.; Jorquin, R.; Braham, J. E. Food
    Tech. 1964, 18, 1599-1603.
18. Reber, E. F.; Hopkins, C. F.; Liu, M. M. Gossypol content of
    cottonseed products processed for human consumption. Paper pre-
    sented at the 44th Annual Meeting of the Institute of Food Tech-
    nology, Anaheim, Ca., June 10-13, 1984. Abstract #9, p. 91 in
    Program.
19. Anonymous. "Official and Tentative Methods of the American Oil
    Chemists' Society"; American Oil Chemists' Society: Champaign,
    Illinois, 1973.
20. Meksongske, L. A.; Clawson, A. J.; Smith, F. H. J. Agric. Food
    Chem. 1970, 18, 917.
21. Nomeir, A. A.; Abou-Donia, M. B. J. Amer. Oil Chem. Soc. 1982,
    59, 546.
22. Hendricks, J.D.; Sinnhuber, R. O.; Loveland, P. M.; Pawlowski,
    N. E.; Nixon, J. E. Science 1980, 208, 309.
23. Lee, D. J.; Wales, J. H.; Sinnhuber, R. O. Cancer Res. 1971,
    31, 960.
24. Scarpelli, D. G.; Lee, D. J.; Sinnhuber, R. O.; Chiga, M.
    Cancer Res. 1974, 34, 2984.
25. Anonymous. Fed. Reg. 1978, 43, 43556.
26. Hardcastle, J. E.; Wei, T. F.; Reber, E. F. "Cyclopropenoid
    fatty acid content of food products prepared with cottonseed
    flour"; Paper presented at the 187th Annual Meeting of the
    American Chemical Society, St. Louis, Mo., April, 1984.
27. Hampden, K. A.; Reber, E. F.; Stewart, G. Fed. Proc. 1983, 42,
    1323.
28. Hampden, K. A. Ph.D. Thesis, Texas Woman's University, Texas,
    1983.
29. Hardy, H. L.; Goyer, R. A.; Guinee, V. F. "Epidemiology and
    Detection of Lead Toxicity"; MSS Information Corp.: New York,
    1976, pp. 34, 115.
30. Pershagen, G. In "Trace Metals: Exposure and Health Effects";
    Diferrante, E., Ed.; Pergamon Press Inc.: New York, 1979; pp.
    99-103.
31. Liener, I. E. "Toxic Constituents of Plant Foodstuffs";
    Academic Press, Inc.: New York, 1980; pp. 319-21.

RECEIVED December 13, 1985

7

Addition of Soy Proteins to Meat Products

A. W. Kotula and B. W. Berry

Meat Science Research Laboratory, Agricultural Research Service, U.S. Department of Agriculture, Beltsville, MD 20705

Soy proteins are used extensively in meat and meat products by the military, the school lunch program and consumers to save money. Their ultimate acceptability is equally dependent upon the nutritional, chemical, sensory and shelf life changes which occur when they are added. Soy proteins in meat products such as ground beef inhibit rancidity, improve tenderness, increase moisture retention, decrease cooking shrink, fat dispersion during cooking and have no important effect on microbiological condition. Concomittantly, inordinate amounts of added soy protein may cause the meat product to be too soft, exhibit an undesirable flavor and may lead to a decreased PER and a deficiency in B-vitamins and trace minerals. In emulsified meat products, soy protein effectively binds water but does not emulsify fat as well as salt soluble muscle protein. Prudent incorporation of plant proteins can result in an improvement of the quality of the meat product with inconsequential adverse effects.

Over a decade ago the institutional trade, the Department of Defense, the School Lunch Program of the United States Department of Agriculture and consumers became interested in utilizing soy proteins to augment the proteins of meat. The economic benefits associated with the use of soy protein were reflected in the 1974 prices for food groups which were utilized as sources of proteins (Table I). Whereas the average price per kg of utilizable protein from meat and fish was in the neighborhood of $6.6 per kg, soybean flour was $.68 per kg, peanut flour $1.45 per kg, and single cell proteins $.95 per kg.
    If one estimated the price of meat at $2.2 per kg and added 25 percent hydrated soy, when the hydration ratio was 3 parts water to 1 part soy, and the cost of the soy was $1.8 per kg, then the savings due to the use of soy would be $.44 per kg of product

Table I. Comparison of Prices of Utilizable Proteins (1)

Food Source	Col. 1 Price of Source Material ¢ per kg.	Col. 2 Crude Protein Content %	Col. 3 NPU Value %	Col. 4 Utilizable Protein Content (Col. 2 X Col. 3) %	Col. 5 Cost of Utilizable Protein (Col. 1 - Col. 4) $ per kg.
Meats and products					
Pork, boneless carcass	94.6	15.7	84.0	13.2	7.16
Beef, boneless carcass	107.1	19.5	76.7	15.0	7.14
Chicken, mature	81.8	19.0	69.6	13.2	6.20
Frankfurters	118.8	14.2	68.0	9.7	12.25
Gelatin	154.0	85.6	2.5	2.1	73.33
Fish	98.1	18.3	79.5	14.5	6.77
Fish protein concentrate	88.0	80.0	71.7	57.4	1.53
Dairy products					
Milk, whole, fluid	14.74	3.5	81.6	2.9	5.08
Milk, skim, powder	49.28	35.6	79.6	28.3	1.74
Cheddar cheese	114.18	25.0	69.8	17.4	6.56
Whey, dried	19.80	12.7	83.9	10.7	1.85
Whey protein concentrate	26.40	75.0	84.0	63.0	0.42
Casein	132.00	99.0	72.1	71.4	1.85
Eggs, medium size	55.00	12.8	93.5	12.0	4.58

Table I continued.

Food Source	Col. 1 Price of Source Material ¢ per kg.	Col. 2 Crude Protein Content %	Col. 3 NPU Value %	Col. 4 Utilizable Protein Content (Col. 2 X Col. 3) %	Col. 5 Cost of Utilizable Protein (Col. 1 ÷ Col. 4) $ per kg.
Legumes and Oilseeds					
Beans, average	14.74	21.4	38.4	8.2	1.80
Peas, dried	12.10	24.0	46.7	11.2	1.08
Peanuts, shelled	40.04	26.9	42.7	11.5	3.48
Soybean flour, low fat	18.70	44.7	61.4	27.4	0.68
Soybeans, extruded	61.60	52.5	58.0	30.4	2.03
Sesame seed	51.92	33.4	53.4	17.8	2.92
Sunflower seed	38.50	23.0	58.1	13.4	2.87
Cottonseed meal or deglanded flour	26.6	42.3	52.7	22.3	1.28
Grains					
Corn meal, whole	14.08	9.2	51.1	4.7	3.00
Flour, white, wheat	14.30	11.8	45.6	5.4	2.65
Flour, white, wheat with 0.3% L-Lysine HCI	15.84	11.8	59.0	7.0	2.26
Wheat gluten	48.62	80.0	37.0	29.6	1.64
Rice, whole	19.80	7.5	70.2	5.3	3.74
Wheat, whole grains	7.26	12.2	65.2	8.0	0.91

(2). Even greater savings, 21 to 35 percent, can be realized
because a certain amount of lean trimmings can be replaced by
fatter trimmings, depending on which formulation is used (Table
II). Augmenting meat or fish protein with soy protein continues to
be economically advantageous. The current price of soy proteins,
(D. R. Rice, Personal communication, 1985), varies with quantities
purchased, but varies on a cent per kg basis from 22 to 55 for soy
flour; 55 to 110 for textured soy flour; 77 to 121 for
non-functional soy concentrate; 132 to 176 for high solubility soy
concentrate; 108 to 154 for textured soy concentrate; 238 to 330
for soy isolate; and 275 to 330 for structured soy isolate.
    Some of the uses of soy proteins to augment meat proteins are
summarized in Table III. Textured soy flour, though not included
in that table, can be used in the same meat systems, where soy grit
or coarse soy protein concentrate are used. Besides the economics
associated with the use of soy protein, they are utilized in such
meat products because of their functional and nutritional
properties.

Functional Properties and Sensory Characteristics

The increased water holding capacity as a result of adding soy
proteins to meat is one of the most important characteristics of
soy proteins. The recommended hydration levels for soy flour (50%
protein), soy concentrate (70% protein) and soy isolate (90%
protein) are about 1.5:1, 2.5:1, and 3.5:1, respectively. Rakosky
(4) recommended using a hydration level which will result in a
protein level of about 18 percent in the hydrated product. Andres
(5) reported that soy concentrate rapidly absorbs and retains over
400% its weight of water. Though the soy concentrate can absorb
even 600 percent of its weight of water, some would be lost due to
syneresis and would detract from the meat product quality. Of
equal importance is the need for the soy protein to retain some of
its water holding capacity during cooking so that it will bind
liberated juices and thus retain the meat flavors and increase the
cooked yield of the meat products (5).
    Rakosky (4) reported that if a 25% shrink were obtained with
an all meat product, the shrink in a meat product augmented with
soy concentrate would be about 22.5%, or a 10% reduction in shrink.
Use of textured soy rather than soy concentrate (Table IV) has been
shown to reduce shrink in beef patties (6,7). In a comparison of
textured soy, soy concentrate and soy isolate, Berry et al. (7)
found isolates gave the lowest cooked yields. In that study,
cooking yields were directly related to the amount of dry soy in
the ground beef-soy formulations. The greater quantity of soy (and
correspondingly less water) in the textured soy flour formulation
increased moisture retention and decreased total cooking loss.
Wolf (8) attributed the water binding by soy flour not only to the
presence of the proteins but also to the polysaccharides which are
present. Kotula and Rough (6) demonstrated that ground beef
patties extended with soy flour or concentrate were more tender
than the all beef patties. Berry et al. (7) found patties made
with soy flour or concentrate to be more tender than all-beef
patties or patties formulated with soy isolate. The dilution

Table II.   Cost Comparison of Ground Beef Blends

	All-Meat		15% Textured Soy Flour (Flake)		20% Soy Protein Concentrate (Grit)		25% Textured Soy Protein Concentrate (Meat-Like Flake)	
	kg.	$Cost	kg.	$Cost	kg.	$Cost	kg.	$Cost
Beef Trim (85/15)	32.5	94.96	22.7	66.5	19.5	57.19	16.4	47.88
Beef Trim (50/50)	13.0	15.73	15.9	19.25	16.8	20.36	17.7	21.45
Protein	-	-	2.3	1.35	2.3	1.95	2.8	3.13
Water	-	-	4.5	-	6.8	-	8.5	-
Total kg.	45.5		45.5		45.5		45.5	
Cost/45.5 kg.		110.69		87.10		79.50		72.46
Cost Difference From All Meat		-		23.59		31.19		38.23
% Approximate Cook Yield	70		78		78		80	
20% Mark-Up		$132.83		$104.52		$95.40		$86.95
Cost Difference				28.31		37.43		45.88

Ingredient Pricing: Beef (85/15) $2.92 kg.
                     Beef (50/50) $1.21 kg.
                     Textured Soy Flour $0.59 kg.
                     Soy Protein Concentrate $0.85 kg.
                     Textured Soy Protein Concentrate $1.12 kg.

Source: Reproduced with permission from reference 3.

Table III. Basic Soy Protein Products Used in Various Meat Systems[a]

Meat system	50% Protein		70% Protein dry basis Soy protein concentrate		90% Protein dry basis Isolated soy protein
	Soy flour	Soy grits	Coarse	Fine	
Ground (coarse)					
Patties	B	A	A	B	B
Meat balls	B	A	A	B	B
Meat loaves	B	A	A	B	B
Chili	A	A	A	B	B
Sloppy joe	B	A	A	B	B
Tacos	A	A	A	B	B
Salisbury steak	A	A	A	B	B
Sausage	AB	A	A	A	A
Emulsion					
Sausage	A			A	A
Bologna	A			A	A
Loaves	A			A	A
Canned	A			A	A
Other					
Baby foods	A	A	A	A	A
Soups					
Canned	A		A	A	A
Dry			A	A	A
Sauces and gravies	A	A	A	A	A
Pet foods	A	A	A	A	A
Poultry rolls	B		B	B	B

[a] A = major additive, B = minor additive.

Source: Reproduced with permission from reference 4.

Table IV. Cooking Characteristics of Beef-Soy Patties

Product	Internal temperature (°C)	S.T.E. Puncture (g)	Shear (g)	Shrink Weight (%)	Diameter (%)	Thickness (%)	Composition of cooking juices Protein (%)	Fat (%)	Moisture (%)
Study I[x]									
All-beef	64.9a	497a	4412a	34.4a	14.9a	8.0a	2.7c	19.3c	77.3b
Textured soy 20%	63.8b	324cd	3076d	22.7c	9.9b	0 c	3.8a	37.4b	57.5c
Textured soy 30%	62.6c	314d	2969e	15.4d	6.8c	3.3b	3.2b	50.3a	45.2d
Soy concentrate 20%	62.7c	388b	3889b	25.3b	9.6b	4.9b	3.1b	11.7d	84.1a
Soy concentrate 30%	63.9b	352c	3611c	24.6b	7.7c	2.9b	3.2b	13.3cd	82.7ab
Study II[z]									
All-beef	--	--	--	44.5a	24.1a	--	--	--	--
Textured soy 20%	--	--	--	36.8c	20.9bc	--	--	--	--
Soy concentrate 20%	--	--	--	39.1b	19.4c	--	--	--	--
Soy isolate 20%	--	--	--	43.7a	22.3ab	--	--	--	--

xFrom (6). Basic values based on 36 observations; Values for composition of cooking juices based on 12 observations because data from 3 patties were pooled in each instance. Means within a column having the same letter or letters are not significantly different (P>.05) according to the Duncan's multiple range test. Three of each type of patty were cooked at 121°C for 8, 9, 10, 11 min.; 149°C for 5, 6, 7, 8 min.; 177°C for 4, 5, 6, 7 min.

zFrom (7). Products contained 20% rehydrated soy for each soy extended product. For this substitution level, 35.7, 27.8 and 22.2% dry soy was in the rehydrated soy for textured soy, soy concentrate and soy isolate, respectively. Means within a column having the same letter are not significantly different (P>.05) according to Duncan's multiple range test.

effect of substituting soy protein for muscle and connective tissue are probably responsible for these tenderness improvements as well as the fact that soluble proteins add to binding properties during the denaturing of proteins associated with cooking. Toasted soy flour, soy grits and certain soy protein concentrates are already denatured prior to their addition to meat.

An evaluation of the composition of the cooking juices, as presented in Table IV, demonstrates, by difference, that fat is retained to a greater extent by the soy concentrate meat product than by the soy flour meat product or the ground beef. The meat product containing soy flour lost more fat during cooking than did the all-beef (Table IV). Similar results were reported by Anderson and Lind (9). When soy protein concentrates are used in canned meat products like chili, the fat islands within the chopped meat products and the fat cap are eliminated (10). When 4% soy concentrate was added to a minced pork product, cook out of fat and moisture was reduced 31% for pasteurized product and 34% for sterilized product.

The stability of emulsified products containing soy protein flour, concentrate or isolate was better than similar emulsions without soy (11). Emulsion stability was higher in emulsions containing soy flour than in those containing soy concentrate or isolate. The use of soy protein isolate has been recommended for non-standardized sausage type products because it binds fat and water, provides cohesiveness, prevents fat migration during or after cooking, adds protein and forms a firm gel structure (12).

In the study by Thompson, et al. (11), the ml of gel released per 100 g emulsion for the reference emulsion without soy, with soy isolate (SIF), soy concentrate (SCF) or soy flour (SF) was 6.07, 5.83, 5.49 and 3.08, respectively, when the hydration ratios were 1:4 (flour:water) for SIF, 1:3 for SCF and 1:2 for SF. The ml gel released per 100 g emulsion containing 10, 15, 20, and 25% soy protein was 6.70, 5.01, 3.94 and 3.57, respectively. When soy protein concentrate was incorporated into an emulsion at the 3.5% level, the processing yields, textural profile and sensory textural attributes of frankfurters were not different among the products with and without added soy concentrate (13). An objective measure of compression and shear modulus indicated that soy protein concentrate incorporated into frankfurters at the 3.5% level had no effect on batter strength or texture (14). The addition of a cottonseed protein to frankfurters to replace 5, 10 or 15% of the meat resulted in higher pH, less cured color, less firmness of skin, softer texture and reduced desirability as judged by a sensory panel (15).

When structured soy protein fiber was added to fermented salami at 15 or 30% levels, trained sensory panels found the flavor to be undesirable, whereas a 116-member untrained panel found the product containing 30% soy flour to be undesirable in flavor, tenderness and overall desirability (16). The flavor of beef patties containing 20% soy protein flour or concentrate was rated about equal to all beef patties by a 52-member panel, whereas patties containing 30% were scored lower by the panel (6). Berry et al. (7) found the characteristic "soy-like" flavor to be more prevalent in patties made with soy flour vs soy isolate and soy concentrate. Compared to all-beef patties, use of soy flour reduced the incidence of rancid flavor, which was probably due in part to the antioxidant properties inherent to soy flour. Fortification of the soy flour with iron and zinc elevated the incidence of rancid flavor.

Nutritional Characteristics

Typical chemical analyses for soy flour and soy concentrate are presented in Table V. Nutritionally, soy proteins are a good source of amino acids, but they could benefit from increased levels of methionine (18). Though such previous literature reported soy proteins to be deficient in methionine, the report of the 1981 Expert Committee on Protein-Energy Requirements, Foreign Agriculture Organization of the United Nations, which is in press, indicated soy protein products meet the requirements for total sulfur amino acids for all age groups except infants. Regardless, methionine is an essential amino acid and protein quality of soy products will be enhanced by increased levels of that amino acid. When Vemury et al. (19) studied the comparative value of several vegetable protein products fed at equal nitrogen levels to human adults, they reported the products containing the highest to the lowest protein nutritional values were egg, beef, blended wheat protein product, extruded soy flour and extruded soy concentrate. The authors also recognized that their use of a nitrogen conversion factor of 6.25% rather than 5.71% for soy protein resulted in a comparison being made between 22.8 g of protein from soy protein with 25 g protein from beef.

The blood cholesterol values of human adults fed test diets of defatted soy flour or extruded soy concentrate tended to be lower than the values for the individuals on the non-test or animal product (beef, egg) diet. The mean value for cholesterol and range, in parenthesis, for the non-test, beef, eggs, soy flour, soy concentrate and blended wheat protein diets in mg/100 mg were 209 (181-250); 200.5 (104-238); 206.8 (164-267); 180.4 (149-216); 191.3 (164-238); and 178.6 (143-192), respectively. Rhee and Smith (20) indicated the cholesterol content of ground beef augmented with soy flour decreased as the amount of added soy increased, due to the dilution of meat which contains cholesterol with soy, which does not. However, in the cooked patties, the amounts of cholesterol were not significantly different from the all beef patties containing 8 or 16% fat, but in the case of the patties containing 27% fat, the patties containing soy protein contained significantly higher levels of cholesterol.

Some legumes, including raw soy or peanut flour are known to contain certain antinutritional factors such as proteinase inhibitors and hemagglutinins or lectins (21,22). These factors can be inactivated, for the most part, by moist heat, during processing. Interestingly, peanut flour contained more trypsin inhibitor and lectin than did soy flour (22).

The controversy concerning the alleged inhibitory activity of soy products to non-heme iron has not yet been resolved completely. Rodriguez et al. (23) discussed the presence of phytate in soy

Table V. Typical Chemical Analysis of Soy Protein Flour and Concentrate (Central Soya, 1985)

CHEMICAL ANALYSIS	FLOUR	CONC
Moisture	6.5%	5.6%
Protein (As is Basis)	53.0%	66.3%
Protein (Moisture-Free Basis)	56.7%	70.2%
Fat	1.0%	0.7%
Ash	6.0%	6.3%
Fiber	3.0%	3.7%
Carbohydrate	31.0%	17.4%
HEAVY METAL (PPM) Maximum		
Mercury	0.05	
Lead	<.05	
Arsenic	<.05	
NUTRITIONAL DATA*		
Protein Efficiency Ratio (PER)	2.1	2.1
(PER of Casein - 2.5)		
Calories per 100 Grams	340	330

AMINO ACID ANALYSIS g/100 g product	FLOUR	CONC
Arginine*	3.8	5.1
Histidine*	1.4	1.8
Lysine*	3.4	4.2
Tyrosine	1.9	2.6
Tryptophan*	0.7	0.9
Phenylalanine*	2.8	3.4
Cystine	0.9	0.8
Serine	2.7	3.4
Threonine*	2.2	2.8
Leucine*	4.0	5.2
Isoleucine*	2.4	3.1
Valine*	2.6	3.3
Glutamic Acid	10.5	13.4
Aspartic Acid	6.4	7.9
Glycine	2.4	2.8
Alanine	2.3	2.8
Proline	2.7	3.9
Methionine*	0.7	0.9

*Essential Amino Acid

*Unfortified product.

Table V Continued:

VITAMIN CONTENT /100 g product	FLOUR	CONC
Vitamin A, I.U.	nil	nil
Vitamin C, mg.	nil	nil
Thiamine (B1), mg.	1.5	0.3
Riboflavin (B2), mg.	0.4	0.1
Niacin, mg.	1.9	0.9
Vitamin B6, mg.	0.5	0.1
Vitamin B12, mcg.	nil	nil
Pantothenic Acid, mg.	1.7	0.7

MINERAL CONTENT mg/100 g product	FLOUR	CONC
Calcium	270	350
Iron	8	10
Phosphorus	730	810
Sodium	20	50
Potassium	2500	2200
Magnesium	300	320
Zinc	5	3

BACTERIOLOGY

Total Plate Count	Less than 15,000 per gram
Coliforms	Less than 10 per gram
Enterk Pathogens	
E. coli	Negative in 0.1 gram
Salmonella	Negative in 200 gram

Source: Reproduced with permission from reference 17.

proteins and its inhibition of iron and other trace mineral bioavailability. They found that iron supplementation of soy protein isolate-based-diet was only slightly effective, thus indicating that added iron may have been bound by the protein or the protein-phytate complex. Treatments of soy involving heating at 120°C for 20 min or reducing the phytate by 75% were equally effective in increasing the bioavailabity of endogenous (39%) or added iron (46%), according to Rodriguez et al. (23). Their studies were carried out with Leghorn cockerel chicks. Other scientists using rats (24), reported the iron bioavailability of soy concentrate to be 92% and soy flour, 85%. Ranger and Neale (25) reported the iron availability, compared to ferrous sulfate (55%), of the various soy proteins to range from 29-57%, for meat samples from 32-39% and for soy meat combination products from 61 to 92%. Those authors noted that in comparison with human studies, the soy iron availability in animal models seems to be overestimated and the meat iron seems to be underestimated using hemoglobin regeneration techniques.

Shelf Life

The typical microbiological quality for soy protein flour and soy protein concentrate were reported (17) as: less than 15,000 per gram for the total plate count, less than 10 per gram for coliforms, Escherichia coli negative in 0.1 g and salmonellae negative in 200 g. During refrigerated storage, aerobic bacteria and coliforms increased faster in soy flour, but not in soy concentrate, than in all beef patties (26). Keeton and Melton (27) observed that aerobic bacteria increased in proportion to the increase in soy. Reports by Judge et al. (28) and Emswiler et al. (29) indicated the addition of soy protein flour or concentrate had no effect of practical significance on the microbial counts. Thompson et al. (30) reported soy significantly increased the rate of growth of total aerobic bacteria, but from his graph there was no difference in rate of growth until after the fourth day of storage at 3°C when the products started to spoil. Harrison et al. (31) reported the shelflife of ground beef with soy could be extended 2 days by the addition of acetic acid. Joseph et al. (32) reported the addition of structured soy protein to the ingredients of dry fermented salami had no effect on the growth of endogenous microorganisms.

The addition of soy protein likewise did not enhance the production of enterotoxin by Staphylococcus aureus S-6 (33). However, the addition of soy protein did increase the pH of the resultant product from 5.8-5.9 to 6.1 and resulted in a significant increase in D-values of salmonellae at 54 and 60°C (34). The addition of soy proteins to beef tended to inhibit rancidity development (35,30,36). Figure 1 graphically demonstrates the inhibitory effect of soy on oxidative rancidity. The addition of soy protein to emulsions had no effect on rancidity (TBA values) (11).

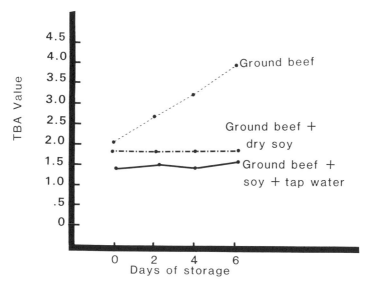

Figure 1.   Rancidity development in ground beef with and without soy
protein. (Reproduced with permission from reference 30.)

Detection

Eldridge and Wolf (37) provided a review of literature for
detecting soy protein by microscopy and histological methods,
immunochemical techniques, electrophoresis, density gradient
centrifugation, fluorescence and chemical determination of unique
components. An additional method published more recently (38)
involved determination of free galactose plus arabinose using
galactose dehydrogenase. Average galactose levels in soy protein
flour, concentrate and isolate were 674, 600 and 66 moles
galactose equivalents per gram of product, respectively. Less than
1 mole of galactose is present per gram of beef or pork. A
histological method for detecting soy at the 3, 5, 10 and 15%
levels was reported by Carey et al. (39). Those authors
successfully tested the procedure using double blind samples
provided to five collaborating laboratories.

Summary

Without a doubt, the greatest incentive for augmenting meat and
meat products with soy proteins is the potential savings in product
costs. Some of the functional properties, such as water holding
capacity, reduced cooking loss and fat dispersion, provide an added
interest in the use of soy proteins. Some of the flavor concerns
expressed over a decade ago are still a reality with some
particular soy products, but for the most part, undesirable flavors
can be overcome by using the recommended lower amounts, or a better
quality soy product or use in meat products containing seasonings.
The soy protein can improve the texture and tenderness of some meat
products. Likewise, the sugars responsible for flatulence can be
removed or greatly reduced. Nutritionally, soy proteins provide a
wide range of amino acids which added to meat appear to be
acceptable for the consuming public. Additional research into
antinutritional factors in soy proteins is indicated. Soy protein
products do aid in inhibiting rancidity and do not contribute
inordinately to the microbial spoilage of the meat products of
which they are a constituent part. Detection methods are presently
available for identifying products which may be mislabeled in
respect to their soy protein content, but additional research
should be carried out to develop methods which are more rapid and
can be carried out by less trained personnel. With the guidelines
of meat safety, proper labeling and good sanitary practices, the
prudent use of soy protein products should provide the meat
industry with greater versatility limited only by their desire to
provide a high quality meat product at a fair price.

Literature Cited

1. Bird, K. Proc. Amer. Inst. Chem. Eng. 1974.
2. Anonymous. Food Proc. 1978, 56-7.
3. Moore, S. Meat Processing 1980, 19, 22, 27-8.
4. Rakosky, J., Jr. J. Amer. Oil Chem. Soc. 1974, 51, 123A-7A.
5. Andres, C. Food Process. 1976, 47-8.

6. Kotula, A. W.; Rough, D. K. Proc. 21st European Meat Res. Work. 1975, p. 168-9.
7. Berry, B. W.; Leddy, K. F.; Bodwell, C. E. J. Food Sci. 1986 (In press).
8. Wolf, W. J. J. Agr. Food Chem. 1970, 18, 969-76.
9. Anderson, R. H.; Lind, K. D. Food Tech. 1975, 44-5.
10. Andres, C. Food Proc. 1983, 74-5.
11. Thompson, L. D.; Janky, D. M.; Arafa, A. S. J. Food Sci. 1984, 49, 1358-62.
12. Anonymous. Food Process. 1975, 92.
13. Keeton, J. T.; Foegeding, E. A.; and Patana-Anaka, C. J. Food Sci. 1984, 49, 1462-5, 1474.
14. Patana-Anake, C.; Foegeding, E. A. J. Food Sci. 1985, 50, 160-4.
15. Terrell, R. N.; Swasdee, R. L.; Wan, P. J.; Lusas, E. W. J. Food Sci. 1981, 46, 845-9.
16. Berry, B. W; Cross, H. R.; Joseph, A. L; Wagner, S. B.; Maga, J. A. J. Food Sci. 1979, 44, 465-74.
17. Central Soya, personal communication.
18. Andres, C. Food Process. 1981, 142-4, 146, 150, 152, 154, 157, 158, 163.
19. Vemury, M.K.D; Kies, C.; Fox, H. M. J. Food Sci. 1976, 41, 1086-91.
20. Rhee, K. S.; Smith, G. C. J. Food Sci. 1983, 48, 268-9.
21. Anonymous. Agricultural Research, USDA 1973, July, 6.
22. Sitren, H. S.; Ahmed, E. M.; George, D. E. J. Food Sci., 50, 418-423.
23. Rodriguez, C. J.; Morr, C. V.; Kunkel, M. E. J. Food Sci. 1985, 50, 1072-5.
24. Picciano, M. F.; Weingarter, K. E.; Erdman, Jr., J. W. J. Food Sci., 1984, 49, 1558-61.
25. Ranger, C. R.; Neale, R. J. J. Food Sci. 1984, 49, 585-9.
26. Craven, S. E.; Mercuri, A. J. J. Food Protect. 1977, 40, 112-5.
27. Keeton, J. T.; Melton, C. C. J. Food Sci. 1978, 43, 1125-9.
28. Judge, M. D.; Haugh, C. G.; Zachariah, G. L.; Parmelec, C. E.; Pyle, R. L. J. Food Sci. 1974, 39, 137.
29. Emswiler, B. S.; Pierson, C. J.; Kotula, A. W.; Cross, H. R. J. Food Sci. 1979, 44, 154-7.
30. Thompson, S. G.; Ockerman, H. W.; Cahill, V. R.; Plimpton, R. F. J. Food Sci. 1978, 43, 289-91.
31. Harrison, M. A.; Draughon, F. A.; Melton, C. C. J. Food Sci. 1983, 48, 825-8.
32. Joseph, A. L.; Berry, B. W.; Wagner, S. B.; Davis, L. A. J. Food Protect., 1978, 41, 881-4.
33. Craven, S. E.; Blankenship, L. C.; Mercuri, A. J. J. Food Protect., 1978, 41, 794-7.
34. Craven, S. E.; Blankenship, L. C. J. Food Protect., 1983, 46, 380-4.
35. Kotula, A. W.; Twigg, G. G.; Young, E.P. Tech Report 75-80-FEL - U.S. Army Natick Research and Development Center, 1974, p. 53.
36. Ray, F. K.; Parrett, N. A.; Van Stavern, B. D.; Ockerman, H. W. J. Food Sci. 1981, 46, 1662-4.

37. Eldridge, A. C.; Wolf, W. J. Proc. World Soybean Res. Conf. II, 1979.
38. Morrissey, P. A.; Olbrantz, K.; Greaser, M. L. Meat Science 1982, 7, 109-16.
39. Carey, A. M., Archer. J. N.; Priore, Jr., J. D.; Kotula, A. W. J. Assoc. Off. Anal. Chem. 1984, 67, 16-9.

RECEIVED March 21, 1986

8

Use of Soy Protein Products in Injected and Absorbed Whole Muscle Meats

L. Steven Young, Greg A. Taylor, and Alexander T. Bonkowski

Archer-Daniels-Midland Company, Chicago, IL 60639

Use of soy protein products in brine injected or absorbed whole muscle meat products such as beef, poultry, and seafood is reviewed. The importance of functionality on brine performance and within muscle tissue is stressed. Major considerations are selection of the proper soy protein, accompanying functionalities such as water-binding, gelling and viscosity, the specific meat system and requirements pertaining to nutrition, processing and marketing. When properly formulated, soy protein allows for significant cost savings, increased yields, reduced fat, increased protein, reduced cholesterol, reduced sodium and/or reduced calories while maintaining muscle tissue integrity. Applicable finished products include ham, roast beef, chicken, turkey, seafood and other whole muscle foods. Finished product character- istics given specific goals and guidelines are out- lined as are new product opportunities.

Given specific goals or guidelines, soy protein products offer versatility and superior functional characteristics for use in a variety of processed meats. When properly selected, soy protein products help reduce finished product costs, maintain quality and create new product opportunities. Exact formulations are determined by muscle product type, cost, quality, nutrition, process, legal and marketing considerations (1-9). Specifically, soy proteins:
1. Increase cooking yield by binding fat and water.
2. Maintain protein content in the finished product when required.
3. Allow use of existing plant facilities.
4. Help utilize available supplies of lean skeletal tissues as well as less desirable pieces or sources.
5. Help reduce finished product costs.
6. Allow precise control of finished product quality.
7. Allow development of new product concepts.

0097-6156/86/0312-0090$06.00/0
© 1986 American Chemical Society

Applicable meat systems include coarse ground (i.e. salami-type), emulsion (i.e. bologna and and weiner-types), restructured (i.e. sectioned and formed) and whole muscle products. The use of soy protein in coarse ground, emulsion and restructured products is well established and understood (3, 5, 7, 9, 10). However, muscle tissue integrity often is sacrificed during product formulation and manufacture. The injection or absorption of salt brines containing soy protein into whole muscle tissue offers opportunity to maintain tissue integrity, take advantage of the unique characteristics of soy protein products, reduce costs, increase yields, and create new products.

Traditionally, brines containing salt(s), phosphate(s), spices, cure, etc., have been successfully used to replace water lost during cooking and, thus, improve yields (5, 6, 9, 11-12). Brines can be injected directly into large whole muscle pieces or gently absorbed into smaller comminuted meat pieces. Cooking yields of less than 100% of green (i.e. starting) weight can be improved to over 100% of green weight. Use of a properly selected soy protein product can allow yields of up to 135% of green weight (4, 5, 7, 9, 11). Until recently, most soy protein technology was directed exclusively toward cured, pumped (i.e. injected) whole pork products such as hams, shoulders, picnics, butts and loins (5, 6, 11). New and existing USDA guidelines regulating meat/non-meat combinations (13) have diversified interest in soy protein injection/absorption technology to other selected muscle tissues (Table I). When properly applied, soy protein brines can offer reduced cost, reduced fat, reduced calorie, portion controlled and/or protein standardized foods with yields up to 200% of green weight while resembling the structure of whole muscle tissue (14).

TABLE I.   WHOLE MUSCLE MEAT SYSTEMS ADAPTABLE TO INJECTION OR
            ABSORPTION PROCESSES

Ham	B-B-Q Ribs
Roast Beef	Shrimp
Corned Beef	Salmon
Turkey Ham	Fish Filets
Chicken Breast	Scallops

Proper selection of soy protein is critical to successful brine formulation and finished product quality (Table II).

TABLE II.   CRITICAL CHARACTERISTICS OF SOY PROTEIN PRODUCTS FOR USE
             IN INJECTION OR ABSORPTION PROCESSES

-Protein Content
-Dispersibility
-Solubility
-Brine Stability
-Brine Viscosity
-Color
-Moisture Retention
-Flavor
-Legal Standards

Soy Protein Products To Consider

Advances in soy protein processing technology have allowed extensive diversification of protein product applications. More sophisticated soy protein products now manufactured have more functionality, better performance, more consistency and better flavor than commercially available defatted soy flour and grits (50% protein dry basis). Among these products are improved textured soy flours, concentrates, and isolates (50%, 70% and 90% protein dry basis, respectfully), functional and non-functional soy protein concentrates (70% protein dry basis) and highly soluble, highly functional isolated soy proteins (90% protein dry basis) (6-8, 14-18).

Textured Soy Proteins. Textured vegetable proteins, primarily textured flours and concentrates (50% protein and 70% protein, dry basis, respectfully) are widely used in the processed meat industry to provide meat-like structure and reduce ingredient costs (3-6, 9-10). Available in a variety of sizes, shapes, colored or uncolored, flavored or unflavored, fortified or unfortified, textured soy proteins can resemble any basic meat ingredient. Beef, pork, seafood and poultry applications are possible (3, 4-7, 15, 19). Proper protein selection and hydration is critical to achieving superior finished product quality. Textured proteins have virtually no solubility and, thus, no ability to penetrate into whole muscle tissue. Therefore, textured soy proteins are inherently restricted to coarse ground (e.g. sausage) or fine emulsion (e.g. weiners and bologna) products, and comminuted and reformed (i.e. restructured) meat products. None are used in whole muscle absorption or injection applications (2-4, 6, 11).

Soy Protein Concentrates. Both non-functional (low or no solubility) and functional (good solubility, emulsification capacity, and dispersibility) soy protein concentrates (70% protein, dry basis) are commercially available for use in meat products (2-4, 6, 9, 15). Normally, a highly functional product with no harsh or bitter flavors is desirable. When used to replace lean meat, non-hydrated concentrate can be used at levels up to 6-7% in finished non-specific emulsion meats. Higher replacement levels or formulas with specific cost/nutrition requirements may use soy protein concentrate with a judicious amount of textured soy protein (6). Excellent yields, cost savings, texture, flavor and nutrient profiles are possible. However, most soy protein concentrates lack sufficient solubility or sufficiently low viscosities to be used in brines for absorption or injection into whole muscle tissue. When legal standards for protein content exist (13), more concentrate must be used to achieve legal minimums. Brine viscosities increase and uniform distribution of brine components throughout the specific whole muscle piece is restricted. Finished product appearance and flavor are easily compromised. Thus, use of soy protein concentrates in whole muscle applications is limited.

Isolated Soy Proteins. Isolated soy proteins (90% protein dry basis) are highly dispersible, highly soluble, highly functional soy products (8, 9, 11, 17, 18). Designed to replace a portion of

salt-soluble meat proteins, isolated soy proteins bind water and
fat, stabilize emulsions and help insure maintenance of structure in
finished cooked meats (3, 5, 6-9). Proper selection is required to
match specific functionality with cost, nutrition, process and marketing needs (Table II). Isolated soy protein can give excellent
emulsification, low viscosity, high moisture retention, and good
stability in high salt concentrations, such as in absorption/
injection brines and within muscle tissue (5-6, 9, 11-12).

Injection/Absorption of Soy Protein Containing Brines

Table I lists whole muscle meats which can be adapted to the injection or absorption of brines containing soy protein. Special considerations and guidelines (9, 12) can be taken into account with
specific goals are defined (Tables III and IV).

TABLE III.   FORMULA CONSIDERATIONS

-Process
-Ingredient
-Nutritional
-Marketing
-Acceptability
-Legal
-Economic

TABLE IV.   NUTRITIONAL GUIDELINES

-Reduced Oil/Fat
-Reduced Cholesterol
-Increased Protein
-Reduced Sweetener
-Reduced Calorie
-Vitamin/Mineral Fortification

Injection Process.   Brines containing soy protein (Table V) may be
injected directly into large, whole muscle pieces such as roast
beef, ham, etc. (5-6, 11-12). Injection is appropriate when tissue
pieces are large, firm, and able to withstand the physical action of
the injection process. Injection is normally accomplished by use of
a mechanical ("stitch") pump. Brine is forced into various areas of
the whole muscle piece through injector needles and distributed
according to needle configuration, size, number and type. Even
distribution of brine within each muscle piece is assured by tumbling or "massaging" injected muscle in a rotating tumbler. Vaccum
may, or may not, be applied to improve brine component distribution
throughout the tissue. Injected and tumbled product can then be
appropriately cooked, cooled, and packaged. Finished products
exhibit whole muscle features including appearance, flavor, color
and slicing characteristics (6, 9, 11-12). In addition, significant
cost savings are achieved while satisfying U.S.D.A. guidelines (13).

TABLE V. Recommended Brine Formulas for 50% Brine Injection Into
          Selected Whole Muscle Tissues

Ingredient	Ham	Corned Beef	Roast Beef
		Percent	
Water	82.2	82.6	86.3
Isolated Soy Protein	7.5	7.5	7.5
Salt	7.0	6.5	5.0
Dextrose	2.0	2.0	--
Sodium Tripolyphosphate	1.2	1.2	1.2
Cure	0.1	0.2	--
Spices/Flavors	as desired	as desired	as desired
Total	100.0	100.0	100.0

Absorption Process. It is possible to absorb brines containing soy
protein (Table VI) into unground, fresh muscle tissue. Most adaptable are small pieces or trimmings with larger available surface
area as compared to large muscle pieces more applicable to injection
processing. This can include pork, beef, mutton, lamb, venison,
turkey, chicken, fish fillets, shrimp, etc. (3, 6, 9, 11-12). Finely ground, comminuted or chopped tissues are not normally compatible
with absorption technology. Absorption can be used to aid the
"restructuring" of pieces which have been reduced in size into a
larger whole muscle structure or to process individually defined
pieces (e.g. fish fillets, shrimp) without "restructuring" (6, 9).
    In general, absorption of brines containing soy protein into
muscle tissue is applicable when:
1. Injection is not possible (meat pieces are too small or fragile
   to handle the mechanical stress of the stitch pump).
2. Meat from less desirable portions (including tissue after boning
   and trimming) are to be combined ("restructured") into more
   desirable and uniform products.
3. Prime pieces are needed to conform to various shapes and sizes
   of containers and packaging.
4. Uniform weights and sizes are required by combining ("restructuring") small and large pieces.
5. Specifications require removal of undesirable internal fat,
   gristle, veins, etc., to yield more desirable and nutritionally
   sound finished products.
6. Tenderization of lesser grade whole muscle meat is desirable.
7. Color uniformity is a major concern.
8. Maintenance of whole muscle tissue integrity is required.
    Brine containing soy protein can be easily absorbed into small
muscle pieces by the use of gentle mechanical action (e.g. tumbling). Again, vaccum may, or may not, be applied to the tumbling
process. As brine is absorbed into the muscle tissue, release of
natural myosin gives each piece the capability of being reunited or
"restructured" into a larger whole muscle piece. Absorption of up
to 100% over green weight is possible. Since tissues remain pli-
able, they can easily be molded or packaged into a variety of shapes
as required. Heat processing solidifies myosin exudate and soy
protein, thus, unifying small muscle pieces into a uniform mass
which will not separate during subsequent handling. When formula-

ted properly, finished restructured products can retain up to 100% of absorbed brine after cooking (i.e. no cooking loss) and products resemble, slice and eat like their whole muscle counterparts (3-6 9, 11, 12).

TABLE VI. Recommended Brine Formulas for Brine Absorption into Selected Whole Muscle Tissue

Ingredient	Roast Beef	Breast Chicken	Shrimp
		Percent	
Water	82.1	90.7	91.0
Isolated Soy Protein	7.3	5.3	5.0
Salt	4.5	2.5	2.0
Dextrose	4.0	--	--
Sodium Tripolyphosphate	1.1	1.5	2.0
Spice/Flavoring	as desired	as desired	as desired
Total	100.0	100.0	100.0

U.S.D.A. Regulatory Guidelines.   In the U.S., whole muscle meat products (i.e. beef, corned beef, ham, pork loins, etc.), which have been absorbed or injected with soy protein containing brine, must comply with the following guidelines (13):
1. Products must be appropriately labeled and standard USDA labeling approval granted.
2. All other U.S.D.A. regulations (e.g. use of phosphates) must be satisfied.
3. The non-meat protein product (e.g. isolated soy protein) must contain a prescribed vitamin/mineral pre-mix.
4. The non-meat protein product must have biological quality of protein (including amino acids added) of not less than P.E.R. 2.0 (80% of casein) or an essential amino acid content (excluding tryptophan) of no less than 28% of total protein.
5. The finished cooked product must contain at least 17% protein.
6. Finished product must be labeled "Combination _____ Product" (Ham, Roast Beef, Turkey, etc.) with an appropriate statement of minimum percent meat ingredient (e.g. 70% Ham, 70% Roast Beef, etc.). Product which has been sectioned and formed (i.e. restructured) must be identified.

Calculation of Brine Components.   In order to satisfy U.S.D.A. regulations, care must be taken during brine preparation so that finished product protein meets or exceeds 17 percent. The following formula can be used to determine brine composition (isolated soy protein, salt, dextrose, polyphosphate, etc.) (6).

$$\text{Percent Dry Ingredient in Brine} = \frac{M + B}{B} \; x \; Y$$

M = Percent Raw Meat = 100%
Y = Percent Ingredient in Uncooked, Injected Product
B = Percent Injected Brine

Example - Raw meat (21% protein) to be injected to 50% over raw weight. Maintain 17% protein in 70% meat finished/cooked combination using isolated soy protein. Assume isolated soy protein is 92% protein.

17% Protein = (70 x .21) + (Y x .92) Therefore,

Y = Amount isolated soy protein in uncooked,
    pumped product = 2.5 percent
M = 100 percent
B = 50 percent

$$\text{Amount Isolated Soy} \atop \text{Protein in Brine} = \frac{100 + 50}{50} \times 2.5\% = 7.5\%$$

Typical selected brine formulas for injection of roast beef, corned beef and ham are given in Table V. Brines for absorption technologies given in Table VI are calculated similarly and may only differ in minor brine components (i.e. salt, seasonings, etc.).

Economics. Tables VII, VIII, IX and X summarize savings (based on ingredient prices, delivered Chicago, Fall, 1985) and yields possible when applying soy protein, injection, absorption and restructuring technologies to the processing of whole muscle meats. Although significant yield increases and savings are possible, extensive new opportunities can result from the development of new products and new product applications.

TABLE VII. COW TOP ROUND - 50% INJECTION PROCESS

	Control	Injected
Cow Top Round	100.0 lbs.	100.0 lbs.
Brine:		
Water		43.0 lbs.
Isolated Soy Protein		3.4 lbs.
Salt		2.5 lbs.
Sodium Tripolyphosphate		0.7 lbs.
Dextrose		0.3 lbs.
Flavor		0.1 lbs.
Total	100.0 lbs.	150.0 lbs.
Ingredient Cost	$117.00	$120.00
Finished Cooked Weight	88.0 lbs.	132.0 lbs.
Cost Per lb. Finished	$1.33	$0.91
Savings Per lb. Finished	--	$0.42
% Savings	--	35.9%

TABLE VIII.   TURKEY ROLL - 35% ABSORPTION/RESTRUCTURE PROCESS

	Control	Absorbed
Turkey Breast (Boneless)	100.0 lbs.	100.0 lbs.
Brine:		
Water		30.5 lbs.
Isolated Soy Protein		2.3 lbs.
Salt		1.4 lbs.
Sodium Tripolyphosphate		0.5 lbs.
Spice		0.3 lbs.
Total	100.0 lbs.	135.0 lbs.
Ingredient Cost	$222.00	$237.67
Finished Cooked Weight	90.0 lbs.	121.5 lbs.
Cost Per lb. Finished	$2.47	$1.96
Savings Per lb. Finished	--	$0.51
% Savings	--	20.6%

TABLE IX.   BOILED SHRIMP - 25% ABSORPTION PROCESS

	Control	Absorbed
70-90 P&D Raw Shrimp	100.0 lbs.	100.0 lbs.
Brine:		
Water		23.0 lbs.
Isolated Soy Protein		1.3 lbs.
Sodium Tripolyphosphate		0.5 lbs.
Salt		0.2 lbs.
Total	100.0 lbs.	125.0 lbs.
Ingredient Cost	$250.00	$251.83
Finished Cooked Weight	72.0 lbs.	95.0 lbs.
Cost Per lb. Finished	$3.47	$2.65
Savings Per lb. Finished	--	$0.82
% Savings	--	24.0%

TABLE X.   SMOKED CHUM SALMON - 33% INJECTION PROCESS

	Control	Injected
Chum Salmon	100.00 lbs.	100.00 lbs.
Brine:		
Water		29.60 lbs.
Isolated Soy Protein		1.70 lbs.
Salt		0.90 lbs.
Sodium Tripolyphosphate		0.70 lbs.
Sodium Nitrite		0.03 lbs.
Sodium Erythorbate		0.07 lbs.
Total	100.00 lbs.	133.00 lbs.
Ingredient Cost	$175.00	$177.00
Finished Cooked Weight	72.0 lbs.	107.0 lbs.
Cost Per lb. Finished	$2.43	$1.65
Saving Per lb. Finished	--	$0.78
% Savings	--	32.1%

New Product Opportunities

Injection and absorption of brines containing soy protein into whole muscle meats can be used to develop a variety of new product concepts designed to meet specific formula guidelines (Table III). Through simple brine reformulation finished meats can be made to comply with desired or legal requirements. Possible are whole muscle foods with reduced fat, reduced cholesterol, reduced sodium, increased protein, reduced simple sugars (i.e. sweeteners) and/or reduced calories (3, 6, 12, 14). Finished products such as these can be used as direct consumer goods or as ingredients in other engineered systems. These, in turn, may include portion and nutrient controlled meals or diets, frozen entrees, sandwiches, etc. Thus, by properly applying the proper soy protein ingredient and brine injection/absorption technologies to control of costs, yields, and nutrient profiles, new product concepts are possible while maintaining the structure, function and integrity of whole muscle tissues.

LITERATURE CITED

1. Brown, W. L. J Am Oil Chemists Soc., 1979, 56 (3), 316-319.
2. Inglett, M. J.; Inglett, G. E. Food Products Formulary, Avi Publishing Co., Inc., Westport, CT, 1982, Volume 4, 89-101.
3. "Ingredient Update 1985", Archer-Daniels-Midland Company, Chicago, IL, 1985.
4. Jul, M. J Am Oil Chemists Soc., 1979, 56 (3), 313-315.
5. Long, L.; Komarik, S. L.; Tressler, D. K. Food Products Formulary, Avi Publishing Company, Inc., Westport, CT, Volume 1, 130-183.
6. "Meat Products Update", Archer-Daniels-Midland Company, Chicago, IL, 1983.
7. Nowacki, J. A. J Am Oil Chemists Soc., 1979, 156 (3), 328-329.
8. Waggle, D. H.; Decker, C. D.; Kolar, C. W. J AM Oil Chemists Soc., 1981, 58 (3), 341-342.
9. Young, L. S. "Soy Protein Products in Processed Meat and Dairy Foods", Presented at World Soybean Research Conference, Ames, IA, 1984.
10. Kadane, V. V. J Am Oil Chemists Soc., 1979, 56 (3), 330-333.
11. Desmyter, E. A.; Wagner, T. J. J Am Oil Chemists Soc., 1979, 56 (3), 334-336.
12. Young, L. S. "Use of Soy Protein Products in Injected and Absorbed Whole Muscle Seafood", Presented at Atlantic Fisheries Technical Conference, Boston, MA, 1985.
13. Federal Register, 105, 21761, 1976.
14. Bonkowski, A., and Taylor, G., unpublished data.
15. Campbell, M. F. J Am Oil Chemists Soc., 1981, 58 (3), 259-261.
16. Kinsella, J. E. J Am Oil Chemists Soc., 1979, 56 (3), 259-261.
17. Ohren, J. A. J Am Oil Chemists Soc., 1981, 58 (3), 333-335.
18. Wilke, H. L.; Waggle, D. H.; Kolar, C. F. J Am Oil Chemists Soc., 1979, 56 (3), 259-261.
19. Sipos, E. F.; Endres, J. G.; Tybor, P. T.; Nakajima, Y. J Am Oil Chemists Soc., 1979, 56 (3), 320-327.

RECEIVED February 10, 1986

BIOLOGICAL EFFECTS

9

Effect of Dietary Protein On Skeletal Integrity in Young Rats

Faustina Bohannon and Gur Ranhotra

Nutrition Research, American Institute of Baking, Manhattan, KS 66502

A number of nutrients affect bone integrity early in life. While the role of certain minerals and vitamins bearing on skeletal integrity is well established, that of protein remains controversial, especially when consumed in excessive amounts. Protein-included calciuric effect as observed in adult man and animals may also occur early in life and thus conceivably affect peak bone mass adversely, particularly when calcium intakes may be marginal. In studies reported here (test model: young female rats), it was found that a diet approaching adequacy in protein and based equally on plant and animal sources would favor some parameters which bear on skeletal mass at maturity more than other combinations of protein consumed.

Several factors, diet-related and others, affect bone mass, and thus skeletal integrity early in life. It is suspected that individuals with large bone mass acquired early in life (peak bone mass) are less likely to develop osteoporosis--thinning of bones--in later years than individuals with lesser bone mass (1-3). While the dietary inadequacy of certain minerals and vitamins that affect skeletal maturity adversely is well documented, for protein our concern centers around the excessive intake of this nutrient. Protein-induced calciuric effect observed in adult man and animals is implicated by some (4-6) as a contributory factor in the etiology of osteoporosis. Protein-induced calciuria, if occurring early in life, may contribute to an early onset of osteoporosis by adversely affecting the peak bone mass, especially when calcium intake may be marginal. The studies reported here were undertaken to examine this latter possibility under conditions wherein the source of dietary protein also varied.

Materials and Methods

Six protein sources--two plant-derived and four animal-derived--were used in this 2 x 3 factorial type study (Table I). Diets were formu-

0097-6156/86/0312-0100$06.00/0
© 1986 American Chemical Society

lated using three combinations of protein sources, to contain 10% or 30% protein. Diets were complete in all nutrients required by the rat (7) except calcium which was provided at 60% of the requirement level. Calcium (0.3%), phosphorus (0.4%), magnesium, zinc, sodium and potassium levels were all equalized between diets. Additional information on diets is provided elsewhere (8).

Table I. Experimental Design (2 x 3 Factorial)

Diet No.	Dietary Protein		
	Level (%)	Type (% of Total)	
		Plant[a]	Animal[b]
A	10	67	33
B	10	50	50
C	10	33	67
D	30	67	33
E	30	50	50
F	30	33	67

[a]As bread: Bread was made with white wheat flour, wheat gluten and defatted soy flour.
[b]Sources used: Cooked ground beef, parmesan cheese, non-fat dry milk and egg white powder.

Weanling female rats were used as the test model in these studies. They were housed individually (10 rats/diet) and offered diet and deionized water ad libitum for 5 months. Apparent calcium and phosphorus absorption and urinary Ca and P losses were measured on collections made the last five days each month. Urine volume and pH values were also recorded. Other data (growth response, serum Ca and P levels, femur mineral composition, femur strength and density, and femur histology) were obtained at the end of the 5-month feeding study. Details of this and the analytical methods used are presented elsewhere (8). All data were subjected to appropriate statistical analyses.

Results and Discussion

Physiological Responses. The various physiological responses measured in rats at the end of the 5-month feeding period are summarized in Table II. These responses were obtained on diets which were complete in all required nutrients except protein and calcium. Protein was provided at a marginal (10%) or excessive (30%) level while calcium was provided at a submarginal level. These levels represent patterns of intake typical of the American population.

With the exception of diet B (10% protein diet based equally on plant and animal protein sources; Table I), the body weight gains of rats (Table II) were significantly higher (Table III) when fed 30% rather than 10% protein diets. Weight gains on diet B equalled or approached gains observed on 30% protein diets (diets D-F). Diet B also showed the highest diet utilization efficiency as the diet: gain ratios suggest (Table II). Irrespective of the protein sources

Table II. Physiological Responses Of Rats (5-Month Experiment)[a]

Parameter	Diets					
	A	B	C	D	E	F
Body wt. gain (g)	214±17	233±4	205±14	233±22	245±23	240±23
Diet intake (g)	1960±97	2003±52	2026±67	2034±105	2130±80	2111±94
Diet:gain (ratio)	9.2±0.6	8.6±0.3	9.9±0.5	8.7±0.4	8.8±0.8	8.8±0.5
Protein intake (g)	196±10	200±5	203±7	610±32	639±24	633±28
Kidney wt. (g/100 g. wt. gain)	0.38±0.02	0.37±0.03	0.40±0.05	0.45±0.04	0.42±0.04	0.42±0.04
Calcium intake (mg)	5880±291	6008±155	6078±200	6102±316	6389±240	6333±281
Phosphorus intake (mg)	7840±388	8011±207	8104±266	8134±420	8519±320	8444±375
Serum Ca (mg/dl)	9.8±0.3	9.9±0.4	9.9±0.3	10.1±0.4	10.1±0.3	9.9±0.2
Serum P (mg/dl)	3.2±0.6	3.3±0.3	3.1±0.3	3.3±0.3	3.4±0.7	3.3±0.5
Femur dry wt.[b] (g)	0.45±0.02	0.48±0.04	0.46±0.03	0.48±0.04	0.50±0.03	0.52±0.04
Femur ash[b] (%)	70.6±0.3	69.9±0.6	70.2±0.5	70.2±0.8	69.9±0.3	70.4±0.6
Femur Ca[b] (%)	26.5±1.3	25.1±2.0	25.7±0.6	26.0±0.4	25.1±1.0	24.9±0.4
Femur P[b] (%)	12.5±0.1	12.4±0.5	12.9±0.7	10.7±0.6	10.7±0.8	10.9±0.4

[a]Values show average (9-10 rats per diet)± standard deviation.
[b]Fat-free, moisture-free basis.

used, high protein diets-induced significant enlargement of the kidney (Tables II and III).

Table III. Statistical Analysis: F Values For
Parameters Included In Table II[a]

	Level	Dietary Protein Type	Level x Type
Body wt. gain	20.2^{***}	4.3^{*}	1.8^{ns}
Diet intake	18.6^{***}	4.6^{**}	0.6^{ns}
Diet:gain	10.5^{**}	7.2^{**}	6.0^{**}
Protein intake	62.9^{***}	3.9^{*}	1.8^{ns}
Kidney wt.	27.2^{***}	0.8^{ns}	3.2^{*}
Ca intake	18.6^{***}	4.6^{*}	0.6^{ns}
P intake	18.6^{***}	4.6^{**}	0.6^{ns}
Serum Ca	4.7^{ns}	0.8^{ns}	1.2^{ns}
Serum P	0.5^{ns}	0.5^{ns}	0.1^{ns}
Femur wt.	18.3^{***}	3.4^{*}	2.0^{ns}
Femur ash	0.6^{ns}	3.6^{*}	2.7^{ns}
Femur Ca	2.3^{ns}	6.6^{**}	0.6^{ns}
Femur P	143.4^{***}	2.3^{ns}	0.3^{ns}

[a] ns, not significant
*, $P \leqslant 0.05$
**, $P \leqslant 0.01$
***, $P \leqslant 0.001$

All diets contained the same level of calcium and phosphorus. However, the diet intake of rats was significantly higher on the 30% as compared to the 10% protein diets (Tables II and III). Consequently, their calcium and phosphorus intakes were also significantly higher (Table III). This did not appear to significantly affect the serum calcium or phosphorus levels which are effectively regulated through homeostatic control (9). The source of protein also did not significantly affect the serum calcium or phosphorus levels.

Femur weights, apparently a function of body weight, were significantly higher (Tables II and III) in rats fed the 30% rather than 10% protein diets. In contrast, percent femur ash (the level obtained suggested a complete ossification of the cartilagenous mass) was unaffected by the level of dietary protein and only marginally affected by the type (source) of protein consumed (Table III). In spite of this, the mineral makeup of the femur ash appeared to differ appreciably. Percent femur phosphorus, but not percent femur calcium, was significantly lower (Tables II and III) in rats fed the 30% compared to 10% protein diets. The physiological significance of this, as it might bear on skeletal mass later in life, is difficult to assess but this difference in ash makeup likely resulted from an excessive urinary phosphorus loss (discussed later) observed on 30% protein diets compared to 10% protein diets.

<u>Urinary Responses.</u> In agreement with some recent findings (10, 11), the apparent absorption of calcium and phosphorus did not increase significantly as protein in the diet increased (data not shown): pro-

Table IV. Urinary Responses Of Rats[a]

Month[b]		A	B	C	D	E	F
					Diet		
1	Ca (mg)	3±1	4±1	3±1	4±1	4±1	4±1
	P (mg)	17±3	31±9	75±9	82±10	94±8	83±10
	Vol. (ml)	65±8	79±23	61±12	80±28	75±20	77±27
	pH	6.2±0.3	6.2±0.4	6.2±0.2	6.2±0.2	6.2±0.2	6.3±0.2
2	Ca (mg)	14±2	13±2	14±4	26±5	18±4	17±3
	P (mg)	42±8	47±7	52±12	65±10	62±8	64±6
	Vol. (ml)	120±8	123±12	130±16	122±10	125±13	126±15
	pH	6.4±0.5	6.2±0.3	6.3±0.3	6.4±0.8	6.7±0.7	6.5±0.5
3	Ca (mg)	27±10	25±4	22±5	38±13	22±6	25±8
	P (mg)	109±10	159±22	119±13	168±29	160±21	148±19
	Vol. (ml)	143±10	146±17	142±8	141±8	147±5	150±12
	pH	6.1±0.2	6.2±0.4	6.2±0.2	6.1±0.9	6.3±0.6	6.5±0.5
4	Ca (mg)	19±3	21±4	15±3	17±4	19±5	23±6
	P (mg)	43±9	51±10	53±16	93±32	60±12	91±34
	Vol. (ml)	156±6	176±3	164±3	132±11	134±26	144±40
	pH	6.3±0.3	6.3±0.3	6.5±0.3	6.3±0.9	6.8±0.8	6.6±0.7
5	Ca (mg)	17±5	12±6	10±2	9±2	9±1	12±5
	P (mg)	45±12	45±14	58±9	58±9	61±12	78±22
	Vol. (ml)	123±21	117±18	112±8	115±13	121±17	132±23
	pH	6.2±0.7	6.4±0.6	6.5±0.5	6.7±0.6	6.6±0.6	6.7±0.8

[a]Values show average (9–10 rats/diet) ± standard deviation.
[b]Based on urine collection made the last five days each month.

tein type also had no significant effect on calcium and phosphorus absorption. Apparently, the significant increases observed in uri-nary calcium and phosphorus losses as protein in the diet increased from 10 to 30% (Tables IV and V) resulted from increased glomerular filtration and/or reduced renal reabsorption. This mechanism is now widely postulated (4, 12) for calcium losses and could also be true for phosphorus. Since the dietary protein sources tested also affect-ed calciuric and phosphaturic responses significantly (Table V), any interpretation of the effect of level of protein, without regard to the source, could be misleading.

Table V. Statistical Analysis: F Values For
Measurements Included In Table IV[a]

	Protein Level (Pl)	Protein Type (Pt)	Pl x Pt	Month (M)	M x Pl	M x Pt	M x Pl x Pt
Urine Ca	5.2^*	7.1^{**}	2.7^{ns}	223.6^{***}	10.5^{***}	5.6^{***}	7.3^{***}
Urine P	302.7^{***}	13.4^{***}	14.8^{***}	327.5^{***}	5.6^{***}	8.3^{***}	7.9^{***}
Volume	0.8^{ns}	1.5^{ns}	1.9^{ns}	262.2^{***}	14.8^{***}	1.0^{ns}	2.5^{***}
pH	2.8^{ns}	0.5^{ns}	0.2^{ns}	8.0^{***}	0.8^{ns}	0.5^{ns}	1.0^{ns}

[a] ns, not significant
*, $P \leqslant 0.05$
**, $P \leqslant 0.01$
***, $P \leqslant 0.001$

In this study, the protein-induced calciuric effect was rather modest. Such could be the response early in life or it may be due to the natural protein sources, which are reported (13, 14) to be less calciuric than purified protein sources. Proteins differ in their calciuric effect and combining them can render them less calciuric. The relative (to calcium) excess of dietary phosphorus, a hypocal-ciuric agent (14) may have also mitigated pronounced hypercalciuria due to excessive protein.

Urine volume increased steadily during the first three months before levelling off (Table IV). Neither urine volume nor urine pH were significantly affected by the level or type of dietary protein or by their interaction (Tables IV and V). For pH, this could be a masking effect resulting from combining different proteins which individually may appreciably affect urine pH.

Other Responses and Measurements. Femur weight and volume (Table VI), apparently a function of body weight, were significantly higher (Table VII) in rats fed 30% compared to 10% protein diets. However, the parameters bearing on skeletal integrity, namely femur density and femur strength, were not significantly influenced by the level or the type of dietary protein. Histological examination of the humerus (transverse and longitudinal sections) also failed to re-veal a significant effect of protein.

Table VI. Femur Density and Strength
(End of Experiment)

	Diet					
	A	B	C	D	E	F
Wet Wt. (g)	0.53	0.56	0.54	0.58	0.60	0.61
	±0.03	±0.05	±0.03	±0.04	±0.03	±0.04
Volume (cm^3)	0.45	0.48	0.46	0.48	0.50	0.52
	±0.02	±0.04	±0.03	±0.04	±0.03	±0.04
Density (g/cm^3)	0.31	0.32	0.31	0.34	0.36	0.36
	±0.02	±0.03	±0.02	±0.03	±0.02	±0.03
Strength (Kg^b)	12.0	13.2	13.0	13.6	12.2	12.6
	±1.3	±2.6	±1.8	±0.8	±1.2	±1.8

[a]Values show average (9-10 rats/diet) ± standard deviation.
Femur strength was measured on right femur; other measurements were made on left femur.
[b]Weight required to break fresh bone placed vertically in
"Instron Press".

Table VII. Statistical Analysis: F Values
For Measurements Included in Table VI[a]

	Dietary Protein		
	Level	Type	Level x Type
Femur Wt.	24.3^{***}	3.0^{ns}	0.8^{ns}
Femur Volume	28.7^{***}	2.2^{ns}	1.0^{ns}
Femur Density	2.9^{ns}	0.1^{ns}	1.3^{ns}
Femur Strength	0.03^{ns}	0.01^{ns}	3.1^{*}

[a] ns, not significant
*, $P \leqslant 0.05$
***, $P \leqslant 0.001$

Acknowledgments

These studies were supported by the Basic Research Fund, American
Institute of Baking and constitute the Ph.D. thesis, Kansas State
University (1984), of Bohannon.

Literature Cited

1. Matokovic, C.; Kostial, K.; Simonovic, I.; Buzina, R.; Brodarec,
    A.; Nordin, B.E.C. Am. J. Clin. Nutr. 1979, 32, 540.
2. Aviolo, L.V. Fed. Proc. 1981, 40, 2418.
3. Irwin, M.I.; Kienholz, E.W. J. Nutr. 1973, 103, 1019.
4. Kim, Y.; Lindswiler, H.W. J. Nutr. 1979, 109, 1399.
5. Calvo, M.S.; Bell, R.R.; Forbes, R.M. J. Nutr. 1982, 112, 1401.
6. Hegsted, M.; Linkswiler, H.W. J. Nutr. 1981, 111, 244.

7. "Nutrient Requirements of Laboratory Animals", National Academy of Sciences, National Research Council, 1978, 3rd ed.
8. Bohannon, F.; Ranhotra, G. Nutr. Rep. Int. 1985, 31, 1291.
9. Clark, I. Am. J. Physiol. 1969, 217, 871.
10. Graves, K.L.; Wolinsky, I. J. Nutr. 1980, 110, 2420.
11. Zemel, M.B.; Schutte, S.A.; Hegsted, M.; Linkswiler, H.M. J. Nutr. 1981, 111, 545.
12. Lindsay, R.; Hart, D.M.; Forrest, C.; Baird, C. Lancet. 1980, 2, 1151.
13. Schutte, S.A.; Zemel, M.B.; Linkswiler, H.M. J. Nutr. 1980, 110, 305.
14. Whiting, S.J.; Draper, H.H. J. Nutr. 1980, 110, 212.

RECEIVED December 13, 1985

10

Effects of Protein Sources on the Utilization of Trace Minerals in Humans

C. E. Bodwell

Energy and Protein Nutrition Laboratory, Beltsville Human Nutrition Research Center, Agricultural Research Service, U.S. Department of Agriculture, Beltsville, MD 20705

The effects of various potentially deleterious components of plant protein sources, as indicated by the results of studies conducted with humans, are briefly discussed. Examples of contradictory results are given. It is concluded that the effects of fiber, phytic acid and protein source per se have not been unequivocally established. Reasons for the general lack of agreement are discussed. Other components which may affect mineral utilization are noted. The practical implications of consuming specific protein sources at customary dietary intake levels are briefly discussed.

From a global view, plant protein sources (cereals, legumes, oilseeds, vegetables, fruits) have always been the primary source of dietary energy and protein for the majority of the world's population. In addition, in recent years, the use of vegetable protein products as sources of dietary protein, as supplements, and as extenders of more traditional animal protein sources has been increasing markedly in the developed countries. These factors are responsible for our interest in determining the effects of plant protein sources on the utilization of minerals in the human diet.

Concerns have been focused on the various constituents present in plant protein sources or products which may have, for the most part, deleterious effects; numerous recent reviews detailing various aspects have been published (e.g., 1-18). In the current discussion, emphasis is placed on the status of our current knowledge about the effects of fiber, phytic acid and protein source per se.

Effects of Fiber

Results from some recent studies ($\underline{19}$-$\underline{36}$) on the effects
of fiber are summarized in Table I. For the most part,
the results are from multi-day balance studies. However,
Turnland et al. ($\underline{36}$) used the stable isotope fecal
monitoring method to assess zinc utilization and Simpson
et al. ($\underline{24}$) measured iron absorption from a single test
meal.

Utilization of iron was not affected at levels of 9
to 26g bran/day from corn or wheat ($\underline{19}$, $\underline{20}$, $\underline{26}$) or by 16
g of bran/day ($\underline{25}$). Negative balances were observed by
Morris and Ellis ($\underline{21}$) during the first balance period
with daily intakes of 36g bran but not during the second
5-day balance period. However, Simpson et al. ($\underline{24}$) found
a marked inhibition of non-heme iron absorption from a
single test meal which included 6 g of wheat bran. With
a diet which provided 35g/day of NDF (neutral detergent
fiber from bran bread), iron balances were decreased ($\underline{26}$)
compared to the 9 or 22g/day intake levels.

Kelsay and co-workers ($\underline{27}$,$\underline{28}$) observed no effects on
iron balance of including 24g/day of NDF from fruits and
vegetables; however, negative zinc and copper balances
were observed. In a second study ($\underline{29}$), lower zinc, but
not copper, balances were observed at intakes of 25g
NDF/day; lower intakes (10, 18g NDF) had no effect. In
these studies, the effects of oxalates (from spinach) may
have affected mineral balances. In a third study, a
combination of consuming spinach every other day plus a
daily fiber intake of 26g NDF resulted in increased fecal
zinc excretion and negative balances ($\underline{37}$).

In several studies (e.g., $\underline{30}$-$\underline{36}$, Table I), the
effects of including "processed" fiber sources (fiber
isolated from natural sources) have been investigated.
Results have been contradictory; for example, 10gm/day
cellulose decreased zinc balances in one study ($\underline{30}$) but
30 to 40g/day did not affect zinc absorption in another
study ($\underline{36}$). Processed fiber may have effects which are
not indicative of the effect of a similar level of
endogenous fiber from foods. In summary, daily intake
levels of fiber equivalent to 20 to 30g NDF/day from food
sources would not be expected to have long-term
deleterious effects on mineral utilization.

Phytic Acid

Phytic acid has been implicated, in both animal and
human studies, as having a deleterious effect on the
utilization of various minerals and in particular iron
and zinc. With reference to zinc, primarily based on
results from animal studies, a molar ratio of dietary
phytic acid to zinc of 12 to 15 (or greater) has been
suggested as a threshold at which decreases in zinc
utilization may occur ($\underline{9}$). Results from some recent
studies, are summarized in Table II.

Table I.   Effects of Fiber

Source	Fiber Intake	Mineral	Effect	Ref.
Wheat bran	14g per day	Zn	None	(19)
Wheat or corn	26g bran per day	Zn,Fe Cu	None	(20)
Wheat bran	36g of bran per day	Zn, Fe	Negative balances, 1st 5-day period; positive balances, 2nd 5-day period	(21-22)
Wheat bran Basal diet	3.3,10.9,or 18.7g/day 18.5g/day	Zn,Fe	No change in balances with bran added	(23)
Wheat bran	6g of bran per test meal	Fe	Marked inhibition (51,74%) of non-heme iron absorption	(24)
Wheat bran	16g per day of raw bran	Fe,Mg Zn	No effect or increased absorption Decreased absorption	(25)
White bread Coarse-bran bread Fine bran bread Whole meal bread	9g NDF[a] per day 22g,35g NDF per day 22g NDF per day 22g NDF per day	Zn,Fe Cu,Mg	Retentions similar for 9 and 22g NDF diets; only deleterious effect observed was that iron balances decreased on 35g NDF diet	(26)

a)NDF=neutral detergent fiber.

on Mineral Utilization in Humans.

Source	Fiber Intake	Mineral	Effect	Ref.
Fruits, vege- tables	24g NDF per day	Fe Zn Cu	No effect Negative balances Negative balances	(27,28)
Fruits, vege- tables	10,18 25g NDF per day	Zn Cu	Lower balances on 25g fiber level No effect	(29)
Cellu- lose	10g per day	Zn	Negative balances	(30)
Cellu- lose, hemi- cellulose or pectin	14.2g per day	Zn	Lowered balances with hemicellulose	(31)
Cellu- ulose, hemi- cellulose or wheat bran	10 or 20g per day	Zn	Increased fecal losses of zinc	(32)
Hemi- cell- ulose	4.2,14.2, 24.2g/day	Zn	Decreasing balances with increased fiber intake	(33)
Veget- arian diet plus cell- ulose, hemi- cell- ulose or wheat bran	14.7g/day (basal) 20g/day 20g/day 20g/day	Fe, Zn	Fecal iron excretion increased with added hemicellulose; omnivores had decreased balances and vegetarians had less positive balances with added cellulose or hemicellulose	(34,35)
Cell- ulose	30-40 g/day	Zn	Absorption not decreased	(36)

Table II. Effects of Phytic Acid

Source	Intake level of phytic acid	Molar ratio (phytic acid to zinc)
Wheat bran	139mg/meal	-
dephytinized	9mg/meal	-
Soy protein	264mg/meal	-
Meals with primary protein from		
chicken	75mg/meal	5.3^a
+ soy flour	200mg/meal	13.4^a
beef	7mg/meal	1.5^a
+ soy flour	200mg/meal	5.2^a
soybeans	620mg/meal	24.4^a
Diets with 70% of protein from		
textured soy	3.8g/day	22.3
soy isolate	1.7g/day	14.2
or animal protein	0.6g/day	3.4
Animal protein diet	0.81g/day	6.4^a
Soy protein diet	1.38g/day	9.6^a
Sodium phytate	2.0g/meal	-
Sodium phytate (formula diets)	2.34g/day	15
Sodium phytate (added to bread)	~1.5g/day	$~13^a$
Sodium phytate (in breads)	-	-
Normal foods	0.5g/day	4.7
+Na Phytate	1.7g/day	16.3
+Na Phytate	2.9g/day	27.5

a) Calculated from authors' data.

on Mineral Utilization in Humans.

Effect	Reference
Removal of of 93% of phytic acid did not improve absorption of non-heme iron	(24)
Removal of phytic acid from soy did not improve non-heme iron absorption	(38)
No effects of phytic acid on absorption; absorption depended on zinc level	(39)
Significantly decreased iron balances with soy isolate diet; tendency for zinc balances to decrease with increased phytic acid level	(40)
Markedly decreased zinc balances with soy diet	(41)
Iron absorption decreased by 77%	(42)
Decreased zinc absorption by almost 50%	(36)
Zinc and iron balances not affected	(23)
Zinc absorptions 2-fold higher with low vs high phytate bread	(43)
No effect on apparent absorption of iron, manganese or copper; although not statistically significant, absorption of zinc tended to decrease with increased phytic acid intake	(44)

Simpson et al. (24) did not observe increased
non-heme iron absorption when dephytinzed bran was
included in a test meal compared to the non-dephytinized
bran. Removal of phytic acid from soy protein likewise
did not improve non-heme absorption (38). Decreased iron
balances, observed by Bodwell et al. (40) with soy
isolate diets, did not appear to be associated with the
level of phytic acid; a tendency for zinc balances to
decrease with increased phytic acid was observed.
However, even with a molar phytic acid to zinc ratio of
22.3, average balances were positive (Table II). In
contrast, at a much lower level of phytic acid intake,
Cossack and Prasad (41) observed negative zinc balances
with a soy protein diet. In the latter study, the soy
products were "washed" with EDTA; the possibility exists
that some residual EDTA remained which could have
affected mineral utilization.

Gillooly et al. (42) observed that iron absorptions
were markedly decreased by the additon of sodium phytate
to test meals while similarly marked decreases in zinc
absorption have been observed (36,43; Table II).
Contradictory results were recently reported (23,44).
Addition of high levels (1.5 to 2.9g/day) of sodium
phytate did not significantly affect the apparent
absorption of iron, manganese, copper or zinc.

It has generally been presumed and observed
(36,42,43), that purified sodium phytate had deleterious
effects in humans similar to those observed in animal
studies while the effects of indigenous phytic acid have
been questioned. Recent results (23,44; Table II)
indicate that even with purified sodium phytate,
deleterious effects are not always observed. The form of
the phytic acid-mineral-protein complex present (12) may
be the determinant factor as to whether a deleterious
effect is observed.

Protein Source

The effects of plant protein sources on iron and zinc
utilization have been of particular concern and the
subject of the most studies with humans. Accordingly,
the following discussion is focused on studies concerned
with these two minerals.

Iron. In the human diet, two forms of iron exist,
heme iron and non-heme iron (14,17). Heme iron is much
more readily available and appears to be absorbed by the
human regardless of other components of the diet which
may inhibit the absorption of the non-heme iron.
However, it has been estimated that about 80% of the
daily dietary intake of iron is in the form of non-heme
iron and only about 20% from heme iron. Hence, any
product added to the diet which causes a significant
decrease in non-heme iron absorption is of concern.

The belief that soy protein, added to beef, caused a
marked decrease in non-heme iron absorption, led to a

number of studies which are summarized in Table III.
Contradictory results have been obtained; however, this
can partly be explained by more recent observations which
will be discussed below. In 3 studies (45-47), in which
men were fed complete diets for 10 to 82 days,
absorptions of non-heme iron were similar to those
observed with milk-based diets (45, 46) or were
maintained at an adequate level (47). However, in these
studies an excess of ascorbic acid, which enhances
absorption of non-heme iron (17), was provided. In a
study in which a more usual level of ascorbic acid was
provided, a soy isolate diet was observed to result in
significantly lower iron balances than those observed
with textured soy or animal protein diets (40).

In a series of studies, Cook and co-workers
(48,49,50) observed a marked decrease in non-heme iron
absorption when soy protein was added to beef or protein
mixtures or when soy protein was consumed as the primary
protein source in test meals and compared to the effects
of including egg albumen or beef alone in the test meal
(Table III). Hallberg and Rossander (38), however, found
less of an effect when total iron absorption (heme plus
non-heme iron) was considered. Likewise, in a study in
which reconstituted textured soy was used to replace 30%
of the beef, the absorptions of total iron were not
greatly altered (51). In a study by Stekel et al. (52),
replacing part of the beef with soy isolate or adding
hydrated isolate to beef decreased non-heme iron
absorption from 12.4% to 9.2 and 9.3%, respectively. In
a related study, non-heme and total iron absorptions were
low when soy isolate was the major protein source.

In a study conducted at Beltsville, Morris et al.
(53) measured absorptions in men before and after they
had consumed beef or beef extended with 6 different soy
products (Table III) as their primary protein source in
9 of 14 breakfast and evening meals for 6 months. No
consistent differences or changes among the groups were
observed.

In the same Beltsville study, no changes in clinical
parameters or in serum ferritin levels were observed in
the men nor in the women and children participating
and consuming beef patties extended with the various
soy products (54, 55).

Hallberg and Rossander (38) fed 9 or 10 subjects a
basal test meal containing maize, rice and black beans
or the same meal with beef or soy flour added.
Non-heme iron absorptions of 3.2, 8.4 and 4.8% and
total iron absorptions of 0.18, 0.63 and 0.51 mg were
observed for the basal meal, the basal meal plus beef,
and the basal meal plus soy flour, respectively. The
authors concluded that the amount of iron absorbed was
"substantially augmented" by the addition of the soy
flour which was high in iron.

Table III. Effects of Soy Protein on Iron Utilization in Humans.

Description	Results	Ref.
Absorption studies (complete diets)		
Five men fed dried skim milk (DSM), DSM plus soy isolate or soy isolate diets, each for 14-day periods; "adequate" levels of ascorbic acid provided	Absorptions (17-21%) of non-heme iron among diets were equivalent	(45)
Eight men fed DSM or soy protein concentrate diets, each for 10-day periods; 72mg ascobric acid consumed with each meal	Non-heme iron absorptions (28-32%) equivalent	(46)
Six men fed soy concentrate diets for 82 days (7 12-day periods); 75mg ascobric acid consumed with each meal	Non-heme iron absorption of 13% (82 days)	(47)
Balance studies (complete diets)		
Diets based on textured soy, soy isolate or animal protein; each diet fed to 16 or 17 men for 35 days; 60mg ascorbic acid/day	Balance values for soy isolate diets significantly lower than values for textured soy or animal diets	(40)
Absorption studies (single meals)		
Infant food supplements (corn, soy, whey and/or wheat protein blends) given in test meals to 3 or 14 adult men	Non-heme iron absorptions were low (0.57-1.40%) for soy containing mixtures	(48)

In Study I, 15 men consumed semi-synthetic meals with (a) egg albumin, (b) casein, or (c) isolated soy protein; in Study II, semi-purified meals were fed to 10 men in which protein was from (a) egg albumen, (b) soy flour, (c) textured soy flour, or (d) soy isolate; in Study III, 11 men consumed meals containing (a) 100g beef, (b) 100g beef plus 30g unrehydrated textured soy, (c) 70g beef plus 30g unrehydrated textured soy or (d) reference iron + ascorbic acid

As measured by extrinsic radioiron labels, non-heme iron absorptions were: Study I, (a) 2.49%, (b) 2.74, and (c) 0.46%; in Study II, (a) 5.50, (b) 0.97 (c) 1.91, and (d) 0.41%; in Study III (a) 3.20, (b) 1.24, (c) 1.51, and (b) 19.88%   (49)

Added 100 mg ascorbic acid to soy isolate or egg albumen semipurified test meals; 9 adult men

Absorptions of non-heme iron were 0.56 and 3.20, (soy isolate, isolate + ascorbic acid), 5.05 and 10.19 (albumen, albumen + ascorbic acid)   (50)

Absorption studies (complete diets)

Semi-purified test meals, fed to 7 men; contained (a) soy isolate, (b) soy isolate, + 100g beef, (c) egg albumen, or (d) egg albumen + 100g beef

Non-heme iron absorptions of (a) 0.36, (b)1.44, (c)5.94, and (d)7.47%   (50)

Series of studies with test-meals in which the major protein sources were (g protein and source): (a) 14g, beef, (b) 7g, beef, (c) 14g, beef, (d) 7g, beef + 7g, textured soy, (e) 7g, beef + 7g, soy flour, (f) 7g, beef + 7g, dephytinized soy flour + labelled hemoglobin; same subjects only within 2 pairs of studies (c-d, e-f)

Non-heme iron absorptions of (a) 11.2, (b) 8.4, (c) 11.2, (d) 7.2, (e) 5.6, (f) 5.2 and (g) 5.2%; total iron (heme + non-heme) absorptions of (c) 0.47mg, (d) 0.33mg, and (e) 0.28mg; in (g), about 22.5% of labelled heme iron absorbed   (38)

(Continued)

Table III. Effects of Soy Protein on Iron Utilization in Humans (Continued).

Description	Results	Ref.
Eleven men fed meals containing (a) 100g beef, (b) 100g beef plus 30g reconstituted textured soy, or (c) 70g beef plus 30g soy product	Non-heme iron absorptions of (a) 3.2, (b) 1.2 and (c) 1.5%; total iron (heme plus non-heme) absorptions of (a) 0.44, (b) 0.36, and (c) 0.41mg	(51)
In Study I (28 subjects), meals contained (a) 100g beef, (b) 100g beef plus 50g hydrated soy isolate, (c) 50g beef plus 50g hydrated soy isolate; in Study II (29 subjects), meals contained (a) intrinsically labeled chicken, (b) soy isolate plus chicken, or (c) soy isolate	In Study I, absorptions of non-heme iron of (a) 12.4, (b) 9.2 and (c) 9.3. In Study II, non-heme iron absorptions of (a) 25, (b) 26.2, and (c) 2.1%; total iron (heme plus non-heme) absorptions of (a) 0.38, (b) 0.22 and (c) 0.05mg	(52)
Absorptions measured in men before and after consuming beef patties (no soy protein), or patties extended (20% level) with soy concentrate, soy isolate or textured soy (with or without fortification of soy products with iron and zinc) for 9 of 14 weekly morning and evening meals for 6 months.	Among groups, mean initial non-heme iron absorptions varied from 2.4 to 5.5%, mean final values from 5.3 to 7.6%; no consistent differences or changes among groups	(53)
Nine or 10 subjects fed meal (a) basal containing maize, rice, and black beans, or the basal meal with (b) 75g beef added, (c) 15g soy flour added, or (d) iron (ferrous sulfate) added in an amount equivalent to that provided by the soy flour in (c)	Non-heme iron absorptions of (a)3.2, (b)8.4, and (c)4.8; total iron absorptions of (a) 0.18, (b)0.63, (c)0.51 and (d)0.64mg. Soy product "substantially augmented" the amount of non-heme iron absorbed from basal meal	(38)

Some of the apparent discrepancies between studies
(often between studies based on long-term feeding vs
those involving single test meals) can, in part, be
explained by the results of a recent study by Lynch et
al. (56). In this study, the heme-iron and the
non-heme iron were separately labeled with different
iron isotopes. The addition of soy protein to beef
caused a marked decrease in non-heme iron absorption
(5.1 vs 1.9%). However, a marked increase was observed
in the absorption of the heme iron upon inclusion of
the soy protein (increased from 33.1 to 42.1%). The
amount of total iron absorbed was decreased by the
inclusion of soy (0.43 mg instead of 0.56 mg); however,
the decrease was not as marked as would have been
suggested by the results of the earlier studies and
might not be deleterious during long-term consumption.
    Lynch et al. (57) found non-heme iron absorptions,
as measured by single test meals, for black beans,
lentils, mung beans split peas, and whole soybeans to
be low (0.84 to 1.91%). This suggests that many
commonly consumed legumes are poor sources of iron;
whether legumes other than soy may have an "offsetting"
enhancing effect on heme-iron absorption cannot be
predicted.
    Zinc. Some recent studies on the effects of soy
protein on zinc utilizaton are summarized in Table IV.
Young and Janghorbani (44) and Istfan et al. (46)
compared the effects of soy isolate or soy concentrate
and dried skim milk as protein sources in multi-day
feeding periods. Zinc absorptions, measured by fecal
monitoring of the extrinsic label given, were
equivalent and no deleterious effects of soy protein
were observed. In a second study, Istfan et al. (47)
fed egg protein diets for 10 days and then a soy
concentrate diet for 82 days. Zinc absorptions were
not decreased by feeding the soy concentrate diet.
    Sandstrom and Cederblad (39) fed single test meals
of chicken, beef, chicken or beef plus soy flour or
soybeans. The amount of zinc in the test meal affected
absorption (Table IV) but not soy flour per se. Higher
levels of zinc resulted in lowered absorptions.
    Janghorbani et al. (58) fed isonitrogenous diets to
10 subjects for 12 days. Both an intrinsic label
(chicken) and extrinsic labels were used. Zinc
absorptions from an all-chicken diet and from a 50%
chicken-50% soy isolate diet were equivalent. Solomons
et al. (59) fed 5 or 10 subjects diets in which milk,
soy isolate, and beef (or mixtures of these) were
protein sources; for the milk and/or soy diets,
absorptions were similar; "fractional" absorptions from
beef bologna may have been higher than from soy bologna
(Table IV).
    In the absorption study conducted by Istfan et al.
(47), zinc balances were also determined. Mean zinc
balances were positive and serum zinc levels were

Table IV.   Effects of Soy Protein on Zinc Utilization in Humans.

Description	Results	Ref.
Absorption Studies		
Five subjects consumed (14 days) formula diets with protein provided from either (a) dried skim milk (DSM) (b) DSM + soy isolate (50:50, protein basis), or (c) soy isolate; extrinsic zinc label (fecal monitoring method)	Absorptions (37-41%) equivalent among 3 diets	(44)
Eight subjects fed (10 days) formula diets with either (a) DSM or (b) soy protein concentrate as protein source; extrinsic zinc label (fecal monitoring method)	Zinc absorptions were equivalent; mean values of (a)29 and (b)26%	(46)
Six subjects fed formula diets with protein provided by (a) egg (10 days) or soy concentrate (82 days); extrinsic zinc label (fecal monitoring method)	Zinc absorptions not different; mean values of (a)31% and (b) 23%; for seven 12-day periods of soy diet, mean values varied from 19 to 32%	(47)
Six to 11 subjects consumed isonitrogenous test meals with primary protein source being (a) chicken, (b) beef, (c) chicken + soy flour, (d) beef + soy flour, (e) soybeans, (f) chicken + zinc, (g) beef + zinc; extrinsic zinc label (whole body retention measured)	Absorptions were (a) 36.2, (b) 20.4, (c) 24.7, (d) 16.5, (e) 19.6, (f) 24.4, and (g) 15.3%; absorption appeared to depend on zinc intake level	(39)

Ten subjects fed (12 days) isonitrogenous diets with all protein from chicken meat or 50% from chicken and 50% from soy isolate; intrinsic (chicken) and extrinsic labels used (fecal monitoring method)	Absorptions of zinc equivalent between diets with similar zinc intakes; absorptions higher with low zinc intake	(58)
Five or 10 subjects fed (12-14 days) diets with (a) nonfat dried milk, (b) soy isolate, (c) milk & isolate, (d) soy isolate bologna and (e) beef bologna; extrinsic zinc labels used (fecal monitoring method)	Mean absorptions were (a) 41, (b) 34, (c) 41, (d) 30 and (e) 41%; no differences between (a), (b), (c); (e) may have "favored zinc absorption" over (d).	(59)

Balance Studies

Six subjects fed formula diets with protein provided by soy concentrate (82 days)	Mean zinc balances varied (+0.14 to +1.38mg/day) for 7 12-day balances; serum zinc levels constant across 82 days	(47)
Subjects (16 or 17) consumed meals with >70% of protein from (a) textured soy, (b) soy isolate or (c) animal protein sources, each for 35 days	Mean balances for days 15-21 not different; although positive, mean balances for days 29-35 were lower for (a) and (b) compared to (c)	(40)
Five subjects consumed (a) animal protein diet or (b) soy protein (flour, isolate) diet for 3 months; soy products "washed" with EDTA	Mean zinc balances lower and negative, -2.39mg/day for (b); +3.25mg/day for (a); plasma zinc levels lower for (b); possibly deleterious effects of "residual" EDTA apparently not determined	(41)

unchanged across the 82-day period of feeding soy
concentrate. In the study by Bodwell et al. (40) mean
balances were lower for textured soy and soy isolate
diets, compared to an animal protein diet, but were
still positive. Conversely, Cossack and Prasad (41)
observed negative balances when a diet containing soy
isolate plus textured soy was fed for 3 months.
However, as noted above, whether or not the EDTA used
to "wash" the soy products was completely removed is
unknown.

Reasons for Discrepancies

As noted above, disagreement has often been observed
among different studies on the effects of fiber, phytic
acid and protein source on mineral utilization. Some
possible reasons include: (a) estimates of absorption
from single meals (with or without previous consumption
of the same foods used in the test meal which may also
affect results) may not always be equivalent to results
from multi-day balance studies, (b) in balance studies,
the failure to allow sufficient time (e.g., 1-2 weeks
or more) for adaptation may alter the findings, (c)
variations in the compositions of meals or diets,
including mineral levels, between studies may influence
the results obtained, and (d) the persons used as
subjects vary and this may have an affect. In
addition, in the fiber studies, the levels, types, and
particle size of fiber fed have varied widely and
levels of other possibly confounding components (e.g.,
caffeine, tanins, oxalates) may have differed.

Practical Implications

Numerous studies (e.g., 60-63) have evaluated the
nutritional mineral status of vegetarians. Most
consume relatively high levels of fiber and some
probably consume a relatively high level of phytic
acid. Although exceptions occur, in general their
mineral status has been adequate. Obviously,
adaptation occurs; this has been shown clinically
(34,35). It thus seems unlikely that increased intakes
of vegetable protein products pose long term risks for
those accustomed to non-vegetarian diets.

Literature Cited

1. Berner, L.A. and Miller, D.D. JAOCS (In Press).
2. Sandstead, H.H. Am. J. Clin Nutr. 1982, 35, 809.
3. Morck, T.A. and Cook, J.D. Cereal Foods World 1981,
   260, 667.
4. Kelsay, J.L. Cereal Chem. 1981, 58, 2.
5. Kelsay, J.L. Am. J. Clin. Nutr., 1978, 31, 142.

6. Bodwell, C.E. Cereal Foods World, 1983, 23, 343.
7. Bodwell, C.E. and Hopkins, D.T. In "New Protein Foods, Vol. 5, Seed Storage Proteins" (Altschul, A.M., and Wilcke, H.L., Eds.), Academic Press, N.Y., (1985) pp. 221-257.
8. Solomons, N.W. Am. J. Clin. Nutr. 1982, 35, 1048.
9. Davies, N.T. In "Dietary Fiber In Health and Disease" (Vahouny, G.V., and Kritchevsky, D., Eds.), Plenum Pres, N.Y., 1982, pp. 105-116.
10. Smith, J.C., Jr., Morris, E.R. and Ellis, R. In "Zinc Deficiency In Human Subjects" (Prasad, A.S. Cabdar, A.O., Brewer, G.J., Aggett, P.J., Eds.), A.L. Liss, N.Y, 1983, pp. 147-169.
11. Erdman, J.W., Jr. Cereal Chem. 1981, 58, 21.
12. Erdman, J.W., Jr. and Forbes, R.M. JAOCS 1981, 58, 489.
13. Turnland, J.R. Cereal Foods World, 1982, 27, 152.
14. Bothwell, T.H., Clydesdale, F.M., Cook, J.D., Dallman, P.R., Hallberg, L., Van Campen, D. and Wolf, W.J. "The Effects of Cereals and Legumes on Iron Availability," Internatl. Nutritional Anemia Consultative Group, The Nutrition Foundation, Washington, D.C., 44 pages, 1982.
15. Kies, C. (Ed.) "Nutritional Bioavailability of Iron", ACS SYMPOSIUM SERIES 203, American Chemical Society, Washington, D.C., 1982.
16. Inglett, G.E. (Ed). "Nutritional Bioavailability of Zinc," ACS SYMPOSIUM SERIES 210, American Chemical Society, Washington, D.C., 1983.
17. Monsen, E.R., Hallberg, L., Layrisse, M., Hegsted, D.M., Cook, J.D., Mertz, W., and Finch, C.A. Am. J. Clin. Nutr. 1978, 31, 134.
18. Reinhold, J.G. In Reference 15, pp. 143-161.
19. Guthrie, B.E. and Robinson, M.F. Fed. Proc. 1978, 37, 254.
20. Sandstead, H.H., Munoz, J.M. Jacob, R.A., Klevay, L.M., Reck, S.J., Logan, G.M., Jr., Dintzis, F.R., Inglett, G.E., and Shuey, W.C. Am. J. Clin. Nutr. 1978, 31, S180.
21. Morris, E.R. and Ellis, R. In Reference 15, pp. 121-141.
22. Morris, E.R. and Ellis, R. In Reference 16, pp. 159-172.
23. Andersson, H., Navert, B., Bingham, S.A., Englyst, H.N. and Cummings, J.H. Br. J. Nutr. 1983, 50, 503.
24. Simpson, K.M., Morris, E.R. and Cook, J.D. Am. J. Clin. Nutr. 1981, 34, 1469.
25. Sandberg, A.-S., Hasselblad, C. and Hasselblad, K. J. Nutr. 1982, 48, 185.
26. Van Dokkum, W., Wesstra, A. and Schippers, F.A. Br. J. Nutr. 1982, 47, 451.
27. Kelsay, J.L., Behall, K.M., and Prather, E.S. Am. J. Clin. Nutr. 1979, 32, 1876.

28. Kelsay, J.L., Jacob, R.A., and Prather, E.S. Am. J.
    Cln. Nutr. 1979, 32, 2307.
29. Kelsay, J.L., Clark, W.M., Herbst, B.J., and
    Prather, E.S., Fed. Proc. 1979, 38, 767.
30. Ismail-Beigi, F., Reinhold, J.G., Faradji, B., and
    Abadi, P., J. Nutr. 1977, 107, 510.
31. Drews, L.M., Kies, C., and Fox, H.M., Am. J. Clin.
    Nutr. 1979, 32, 1893.
32. Papakyrikos, H., Kies, C., and Fox, H.M., Fed.
    Proc. 1979, 38, 549.
33. Kies, C., Fox, H.M., and Beshgetoor, D., Cereal
    Chem. 1979, 56, 133.
34. Kies, C., Young, E. and McEndree, L. In Reference
    16, pp. 8-126.
35. Kies, C. and McEndree, L. In Refernce 16, pp.
    183-198.
36. Turnland, J.R., King, J.C., Keyes, W.R., Gong, B.
    and Michel, M.C. Am. J. Clin. Nutr. 1984, 40, 1071.
37. Kelsay, J.J. In Reference 16, pp. 127-143.
38. Hallberg, L. and Rossander, L. Am. J. Clin. Nutr.
    1982, 36, 514.
39. Sandstrom, B. and Cederblad, A. Am. J. Clin. Nutr.
    1980, 33, 1778.
40. Bodwell, C.E., Smith, J.C., Judd, J., Steele, P.D.,
    Cottrell, S.L., Schuster, E., Staples, R., XII
    International Nutrition Congress, (Abstr), San
    Diego, CA (1981).
41. Cossack, Z.T. and Prasad, A.S. Nutr. Res. 1983, 3,
    23.
42. Gillooly, M., Bothwell, T.H., Torrance, J.D.,
    MacPhail, A.P., Derman, D.P., Bezwoda, W.R., Mills,
    W. and Carlton, R.W. Br. J. Nutr. 1983, 49, 331.
43. Navert, B., Cedarblad, A. and Sandstrom, B. In
    Reference 25.
44. Morris, E.R. and Ellis, R. 1986. In "Trace Element
    Metabolism In Man And Animals-IV." (Mills, C.F.,
    Aggett, P.J., Bremner, I., Chesters, J.K., Eds.)
    Cambridge University Press, London, 1986 (In
    Press).
45. Young, V.R., and Janghorbani, M. (1981). Cereal
    Chem. 1981, 58, 12.
46. Istfan, N., Murray, E., Janghorbani, M., and Young,
    V.R. J. Nutr. 1983, 113, 2516.
47. Istfan, N., Murray, E., Janghorbani, M., Evans,
    W.J., and Young, V.R. J. Nutr. 1983, 113, 2524.
48. Morck, T.A., Lynch, S.R., and Cook, J.D. Am. J.
    Clin. Nutr. 1981, 34, 2630.
49. Cook, J.D., Morck, T.A., and Lynch, S.R. Am. J.
    Clin. Nutr. 1981, 34, 2622.
50. Morck, T.A., Lynch, S.R., and Cook, J.D. Am. J.
    Clin. Nutr. 1981, 36, 219.
51. Cook, J. Data published in Reference 14.

52. Stekel, A. Data published in Reference 14.
53. Morris, E.R., Bodwell, C.E., Miles, C.W., Mertz, W., Prather, E.S., and Canary, J.J. Fed. Proc. 1983, 42, 530.
54. Miles, C.W., Bodwell, C.E., Morris, E.R., Mertz, W., Canary, J.J., Prather, E.S., Fed. Proc. 1983, 42, 529.
55. Bodwell, C.E., Miles, C.W., Morris, E.R., Mertz, W., Canary, J.J., and Prather, E.S., Fed. Proc. (1983) 42:529.
56. Lynch, S.R., Dassenko, S.A., Morck, T.A., Beard, J.L., and Cook, J.D. Am. J. Clin. Nutr. 1985, 41, 13.
57. Lynch, S.R., Beard, J.L., Dassenko, S.A. and Cook, J.D. Am. J. Clin. Nutr. 1984, 40, 42.
58. Janghorbani, M., Istfan, N.W., Pagounes, J.O., Steinke, F.H., and Young, V.R. Am. J. Clin. Nutr. 1982, 36, 537.
59. Solomons, N.W., Janghorbani, M., Ting, B.T.G., Steinke, F.H., Christensen, M., Bijlani, R., Istfan, N. and Young, V.R. J. Nutr. 1982, 112, 1809.
60. Anderson, B.M., Gibson, R.S. and Sabry, J.H. Am. J. Clin. Nutr. 1981, 34, 1042.
61. Harland, B.F. and Peterson, M. J. Am. Dietet. A. 1978, 72, 259.
62. Dwyer, J.T., Dietz, W.H., Jr., Andrews, E.M. and Suskind, R.M. Am. J. Clin. Nutr. 1982, 35, 204.
63. Schultz, T.D. and Leklem, J.E., J. Am. Dietet. A. 1983, 83, 27.

RECEIVED March 17, 1986

11

Protein–Procyanidin Interaction and Nutritional Quality of Dry Beans

W. E. Artz[1], B. G. Swanson, B. J. Sendzicki, A. Rasyid, and R. E. W. Birch

Food Science and Human Nutrition, Washington State University, Pullman, WA 99164–6330

Thermodynamic analysis of the temperature dependance of procyanidin binding to bovine serum albumin (BSA) and bean glycoprotein G-1 suggested predominantly hydrophobic and hydrophilic binding, respectively. A cis-parinaric acid fluorescence assay for surface hydrophobicity supported amphiphilic interactions of procyanidin. Heat denatured G-1 had a surface hydrophobicity greater than native G-1. Procyanidin dimer and trimer inhibited trypsin digestion of BSA. In vitro digestibility and Tetrahymena-Protein Efficiency Ratio (t-PER) were inversely related to procyanidin concentration. Procyanidin intubation restricts rat growth and damages intestinal villi. Procyanidins intubated with food or as dry beans were not as inhibitory as procyanidins intubated alone. Digestibility and PER of tempeh prepared with red beans and corn were less than the digestibility and PER of soybean tempeh. Tempeh, Rhizopus oligosporus, fermentation did not improve digestibility or nutritional quality of dry black beans.

The common dry bean, Phaseolus vulgaris, is a grain legume consumed in large quantities around the world. Black and other colored beans provide appreciable protein, vitamins, minerals and calories for rural and urban populations of developing countries. The nutritional importance of beans is great since access to protein of animal origin is limited. Legumes and cereals, which contain complementary proteins, provide protein of greater quality than consumption of legumes or cereals alone. However, consumption of beans and cereals in a favorable nutritional quality ratio and amount tends to be infrequent in developing countries. World production of legumes appears to be declining compared to production and greater yields of cereals. Legume production, however, is still encouraged internationally to fix atmospheric nitrogen and contribute to increased soil fertility in developing countries. Dry

[1]Current address: Food Science, University of Illinois, Urbana, IL 61801.

beans are also an excellent source of complex carbohydrates, fibre and polyunsaturated fatty acids. However, dry beans have several undesirable attributes such as enzyme inhibitors, phytates, flatus factors, lectins, allergens and condensed tannins that constrain nutritional quality unless destroyed or removed.

This paper presents research data that delineate the relationship of dry bean proteins to dry bean procyanidins, and discusses the constraints protein-procyanidin interaction places on nutritional quality of dry beans.

Proteins in Legumes

Proteins present in the seeds of legumes are primarily of two types: 1) enzymatic and structural metabolic proteins responsible for normal cellular activities including the synthesis of structural proteins, and 2) storage proteins. The storage proteins and reserves of carbohydrates and oils are synthesized during seed development (1). Storage proteins occur within the cell in discreet protein bodies (Figure 1) that develop late during maturation of bean seeds (2). The quantity of protein in dry beans ranges from 18 to 25% (180 - 250 g/kg) dry weight. Protein fractionation studies of Phaseolus vulgaris L. have generated three major soluble protein fractions: phaseolin (G1), globulin (G2) and albumin (3-4).

Considerable confusion surrounds the nomenclature of seed proteins of common beans. Phaseolin is reportedly the preferred trivial designation for the globulin-1 (G1), glycoprotein II or vicilin, a 6.9S protein which aggregates to form the 18S tetramer at pH 4.5 (5, 1). Globulin-1 (G1), globulin-2 (G2) and albumin represent 36-46%, 5-12% and 12-16% of the total seed protein, respectively, although there appeared to be some contamination between the latter two fractions after usual isolation procedures (4). Environmental factors such as geographic location and growing season substantially influence protein content of dry beans (6). Polypeptide classification of G1 fraction has been well documented and permits classification of bean cultivars into three groups: Tendergreen, Sanilac and Contender, on the basis of electrophoresis banding patterns (7). The major storage protein in beans, globulin-1 (G1), exhibits a pH dependent polymerization that was utilized in purification (8). At pH 3.8 to 5.4, G1 exists as a tetramer, while at pH 6.4 to 10.5, G1 exists as a monomer of MW 163,000 (9). The isoelectric point of G1 is pH 4.4 - 5.6. G1 solubility is independent of pH from pH 2.5 to 12.0 (10) in 0.5F NaCl. A crude extract of G1 was prepared for purification on cyanogen bromide activated Sepharose.

Tannins/Procyanidins

Tannins are one of several antinutritional factors present in dry beans. Any polyphenolic compound that precipitates proteins from an aqueous solution can be regarded as a tannin (11). Tannins precipitate proteins due to functional groups that complex strongly with two or more protein molecules, building up a large cross-linked protein-tannin complex (12).

Naturally-occurring food legume tannins are reported to interact with enzyme and non-enzyme proteins to form complexes that

result in inactivation of digestive enzymes and protein insolubility
(13). In vitro and in vivo studies indicate that bean tannins
decrease protein digestibility and protein quality (14).
Condensed tannins and procyanidins are terms used to describe
the same general class of compounds, plant phenolics that are
polymers of the flavan-3-ols (Figure 2), (+) catechin and/or (-)
epicatechin (15). Procyanidins heated in alcohol and acid will
produce colored compounds structurally related to anthocyanidins
(16). Procyanidin polymers consist of chains of 5, 7, 3', 4'-
tetrahydroxyflavan-3-ol connected by C(4)-C(6) or C(4)-C(8) bonds
(15). Procyanidins occur free and not as glycosides (17). An
additional hydroxy group can sometimes be found on the B ring of the
flavan-3-ol at the 5' position. Some hydroxy groups on the B ring
may be methoxylated (18) and the methoxyl groups may affect
protein-procyanidin interaction. Procyanidin concentrations range
from 1.5 to 18.6 mg of procyanidin per gram of whole bean flour (4).
Procyanidins are found in greater concentrations in colored beans
than in white beans, most of which are located in the seed coat,
testa or hull.

Methods

Dimeric and trimeric catechin were prepared by reduction of
dihydroquercetin with sodium borohydride in the presence of
catechin (19). Polymerization was followed on silica gel TLC using
an acetone:toluene:formic acid (60:30:10) solvent (20). The
visualization agent was vanillin (1 g/100 ml) in 70% v/v sulfuric
acid/water. Catechin, catechin dimer and catechin trimer appeared
as red spots with respective Rf values of 0.66, 0.54 and 0.43. The
flavan-3,4-diol appeared as a purple spot with an Rf value of 0.63.
Separation of dimeric and trimeric catechin was accomplished with
Sephadex LH-20 using an ethanol:water (45:55) solvent. Purity of
the procyanidin was evaluated with reversed-phase HPLC isocratically
with 4% acetic acid in water. Tritium-labelled procyanidin dimer
and trimer were synthesized similarly with 25 mCi tritiated sodium
borohydride added to the reaction mixture over a 20 min period under
nitrogen. Binding constants of ligand, tritium labelled catechin
dimer and trimer, to defatted bovine serum albumin (BSA) and
purified bean protein G1 were determined by the method developed by
Sophianopoulos et al. (23) with an Amicon Micropartition System
MPS-1 (American Corp., Danvers, MA). Separation of the free ligand
from the bound ligand was accomplished by convective filtration of
free ligand through an anisotropic, hydrophilic YMT ultrafiltration
membrane. The driving force was provided by centrifugation.
Proteins were quantitatively retained above the membrane while low
molecular weight ligand passed through the membrane. Binding
constants were determined by Scatchard plot analysis (24-26).
Linear regression analysis was used to fit the points for the
Scatchard plot.

Procyanidin Binding to Bovine Serum Albumin (BSA)

Polymeric procyanidin extraction from black beans (Phaseolus
vulgaris L. cv. Black Turtle Soup) and purification was a

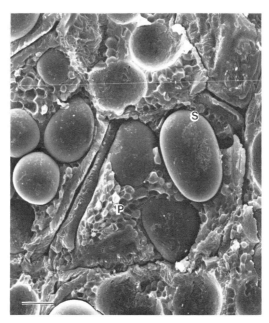

Figure 1.   Scanning electron micrograph of <u>Phaseolus</u> <u>vulgaris</u>
cotyledon showing protein bodies (P) and starch granules (S).
Bar = 10 μm.

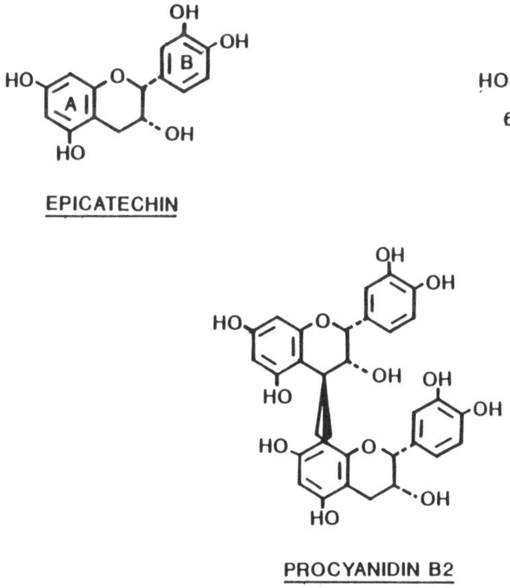

EPICATECHIN   CATECHIN

PROCYANIDIN B2

Figure 2.   Structure of epicatechin, catechin and procyanidin
dimer B2.

modification of the procedure of Strumeyer and Malin (21) and the
procedure of Cansfield et al. (22) using 80:20 ratio of
acetone:water as primary extracting solvent, and LH-20 for clean-up.
The method of Kato and Nakai (27) for determining protein
surface hydrophobicity was adapted for evaluating procyanidin
binding to BSA and G1. The procedure is based on the fact that the
fluorescence quantum yield of cis-parinaric acid increases 40-fold
when cis-parinaric acid enters a hydrophobic environment from a
hydrophilic environment. The digestion of BSA by trypsin in the
presence of procyanidin dimer, procyanidin trimer and black bean
procyanidin polymer was evaluated by discontinuous sodium dodecyl
sulfate (SDS) slab gel electrophoresis and a picryl sulfonic acid
(TNBS) assay (28).

Scatchard plots were used to determine the binding constants of
procyanidins to BSA and bean globulin G1 at temperatures of 19, 29
and 39°C (Figures 3 and 4). Nu (v) is moles of ligand bound per
mole of protein. L is the concentration of the free ligand. The
equilibrium binding constant is equal to the negative slope of the
corrected curve as determined by linear regression analysis from the
Scatchard plot. Nonspecific binding is the binding of ligand to
protein sites possessing low affinity (24). High affinity binding
must be corrected for nonspecific binding (25). Nonspecific binding
was determined as the y-axis intercept of the extension of the lower
affinity binding curve. The lower affinity nonspecific binding was
subtracted from binding possessing the high affinity to produce the
corrected binding curve. The negative slope of the curve is equal
to the equilibrium association binding constant and the x-axis
intercept is equal to the moles of ligand bound per mole of protein.

Thermodynamic analysis of the binding constants of BSA and
procyanidin dimer and trimer from the Van't Hoff equation (29)
indicates a reaction with a positive entropy change, a positive

Table I. Binding Constants

	G1		BSA
			Procyanidin
Temperature	Trimer	Dimer	Trimer
	(k)	(k)	(k)
19	74,000	4,000	110,000
29	42,000	8,200	120,000
39	27,000	20,490	122,000

Table II. Enthalpy, entropy and free energy

Procyanidin	Protein	Enthalpy (H)	Entropy (S)	Free Energy (G)
		(kcal/mole)	(eu)	(kcal/mole)
Trimer	G1	- 9.26	- 9.41	- 6.51
Dimer	BSA	14.9	67.5	- 4.81
Trimer	BSA	0.96	26.3	- 6.73

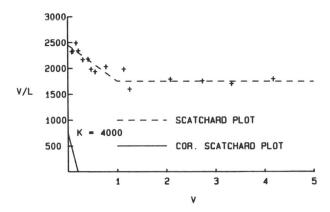

Figure 3.   Scatchard plot of BSA and procyanidin dimer at 19C.

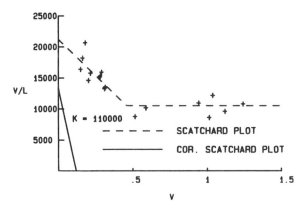

Figure 4.   Scatchard plot of BSA and procyanidin trimer at 19C.

enthalpy change and a negative free energy, i.e. a spontaneous reaction that is totally entropy driven (Tables I and II). Since hydrophobic interactions are entropy driven, the binding of BSA to procyanidin is hydrophobic.

In aqueous solutions, hydrophobic interaction is very important (30). Water molecules at the surface of the hydrophobic domain created by a nonpolar solute rearrange in order to regenerate broken hydrogen bonds, but in doing so create a greater degree of local order than exists in pure liquid water, thereby producing a decrease in entropy (30). The driving force for hydrophobic interaction is the increase in entropy when the ordered water is released to the bulk water. Hydrophobic interactions are entropy driven.

Certain types of non-covalent interactions such as hydrogen bonds, London interactions and van der Waals interactions are enthalpy driven interactions (26); heat is released during bond formation. The heat released during bond formation stabilizes the bonds. Hydrogen bonds, London interactions and van der Waals interactions are variants on the dipole-dipole interaction model, which include permanent and induced dipoles.

Trimeric procyanidin binds more tightly to BSA than dimeric procyanidin (Table II). Partition coefficients of dimeric and trimeric catechin between n-octanol and water indicate procyanidin trimer is more hydrophobic than procyanidin dimer. Increased binding constants of trimer relative to dimer agree with reported partition coefficients. Surface hydrophobicity assays with cis-parinaric acid confirm the thermodynamic analysis that binding of procyanidin to BSA is hydrophobic.

Procyanidin Binding to Bean Globulin (G1)

Binding of procyanidin trimer to bean protein G1 was temperature dependent (Figure 5). An increase in temperature resulted in a large decrease in the binding constant (Table I). G1 binding to procyanidin trimer is spontaneous and hydrophilic in nature. The binding is driven by the large change in enthalpy (Table II). The type of bonding involved between G1, a glycoprotein, and procyanidin is probably hydrogen bonding. Evaluation of the G1 interaction with procyanidin trimer with cis-parinaric acid confirmed that the binding of native G1 to procyanidin trimer is hydrophilic (Figure 6).

Heat-denatured G1 exhibited a surface hydrophobicity greater than that of native G1. The increase was not unexpected since hydrophobic groups are commonly oriented towards the center of proteins in aqueous solvents. Heat denaturation of protein exposes hydrophobic groups to the solvent. Binding of denatured G1 to bean procyanidin oligomer was predominantly hydrophobic.

Common bean procyanidins are capable of both hydrophilic and hydrophobic interaction with protein. Hydrophilic interactions are favored with a hydrophilic glycoprotein like common bean globulin G1, while hydrophobic interactions are favored after protein denaturation, when protein hydrophobic groups are exposed to the solvent.

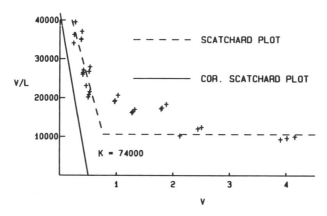

Figure 5. Scatchard plot of procyanidin trimer and bean globulin G1 at 19C.

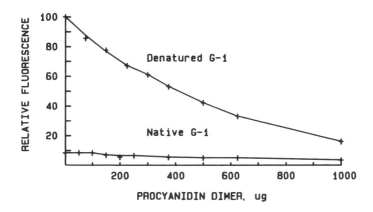

Figure 6. Fluorescence of G1 (0.04%), procyanidin dimer and cis-parinaric acid.

Trypsin Inhibition

Table III. Percent Inhibition of the Trypsin Digestion of BSA

Procyanidin	1 mg/ml	5 mg/ml
Dimer	16.9	25.2
Trimer	42.3	71.2
Polymer	50.5	89.4

Trypsin digestion of BSA was inhibited by addition of procyanidin
dimer, trimer and oligomer (Table III). Increased procyanidin
concentration increased inhibition of the trypsin digestion of BSA.
Increased procyanidin chain length also increased inhibition of
trypsin digestion. Protease inhibition by procyanidin does not
occur by irreversible binding of procyanidin to the active site of
the protease. Procyanidin is not a specific inhibitor for either
trypsin or chymotrypsin, i.e. procyanidin does not inhibit by
binding irreversibly to the active site, rather procyanidin binds
non-specifically to the enzyme and/or protein substrate. Since
procyanidin does not bind specifically to protease active sites, but
reacts non-specifically, Scatchard plots indicate less than one mole
of procyanidin is bound per mole of protein. With polymeric
procyanidin, considerable cross-linking will occur. Not all
anti-nutritional effects can be explained by high affinity binding.

Feeding Procyanidins

Complete removal of procyanidin is not necessary to overcome
anti-nutritional effects. Removal of most procyanidin or addition
of sufficient protein will overcome anti-nutritional effects of
procyanidins. Small concentrations of procyanidin can be easily
overcome by adding protein.
        The most apparent nutritional effect of feeding procyanidins at
naturally occurring concentrations in plants, such as in sorghum
grain (1-2%), are growth depression, poor feed efficiency ratios and
increased fecal nitrogen (12). Protein Efficiency Ratio is a
procedure to measure the ratio of weight gain to protein intake of
weanling rats fed a diet with a single suboptimal 10% concentration
of test protein. Tetrahymena-PER is a more rapid assay, using
protozoa Tetrahymena pyriformis or Tetrahymena thermophila, as an
alternative to the laboratory rat as a biological assay for protein
quality. Good correlation between PER determined with the rat and
Tetrahymena have been reported (14). In vitro digestibility and
Tetrahymena-PER are inversely related to procyanidin content (14).
Bioavailability, expressed as Tetrahymena growth, of bean globulin
G1 in the presence of black bean procyanidins correlated well with
in vitro digestibility of the protein.
        Health consequences of procyanidins in the human diet are
relatively unknown, but the toxicity for human beings may be similar
to the toxicity observed in experimental animals (12). Bender and
Mohammidiha (31) proposed that increased fecal nitrogen from rats
fed diets containing large quantities of cooked legumes was due to
increased gastrointestinal mucosal cell turnover, rather than poor

protein digestibility. Fairweather-Tait et al. (32) discovered that
mucosal cell sloughing increased 35% in the small intestine of rats
fed beans compared to rats fed a control diet. Physiological
alterations such as damage to the mucosal lining of the
gastrointestinal tract and increased cation excretion have also been
demonstrated (33-35). Very great concentrations of dietary
procyanidin, near 5%, can cause death (12). Increased fecal
nitrogen or decreased nitrogen retention by animals fed procyanidin
has been explained by either a reduced digestibility of dietary
protein or an increased excretion of endogenous protein (33).
Explanations for the antinutritional aspects of procyanidins have
centered around the ability of procyanidin to bind protein.
    Rats intubated with 5.0% procyanidin developed coughing,
sneezing, wheezing, overall respiratory distress and severe
dehydration, and were sacrificed after 20 d. Gross pathological
examination revealed moderate to large quantities of intestinal gas,
distended intestinal walls and a translucent quality to the small
intestinal mucosa. The duodenum was discolored, black-purple, for
0.5 to 1.0 cm aboral to the pyloric sphincter. The jejunum and
ileum were thin-walled translucent and gas-filled when compared to
jejunum and ileum of control rats (36). Histological examination of
gastrointestinal tissues from rats intubated with 5% procyanidin
revealed broad, short and fused villi in the areas where the
duodenal tissue was dissolved. Dietary procyanidin can damage villi
decreasing the absorptive surface area and altering the absorptive
capability of the intestinal mucosa. Nutrient
availability is reduced and dietary protein deficiency can result.
Gastrointestinal epithelial damage observed with purified
procyanidin may be dose dependent. Intubations of 1.0 and 0.5%
procyanidin were not toxic over a four week period, yet resulted in
areas of villi shortening and broadening in some of the rats
intubated. Growth rate reduction did occur with 1.0 and 0.5%
procyanidin intubation with food. Long term consumption of
unpurified procyanidins contained in legumes had no detectable
effect on the histological appearance of the gastrointestinal tract
in rats consuming diets prepared with 40% black beans.

Dry Bean Fermentation-Tempeh

Tempeh, an Indonesian food generally produced from soybeans
fermented by Rhizopus oligosporus, is more acceptable than cooked
soybeans because, in part, tempeh does not have the unacceptable
beany flavor and flatus problem associated with soybeans. Tempeh
prepared with small red beans or a small red bean/corn mixture were
acceptable in color, sweeter and more fragrant in flavor and similar
in texture to soybean tempeh. The PER and in vitro protein
digestibility of small red bean (1.69, 85.2) and small red bean/corn
(2.15, 86.1) tempeh were less than the PER and in vitro protein
digestibility of soybean (2.63, 88.9) and soybean/corn (3.11, 90.2)
tempehs (37). Tempeh fermentation does not improve the protein
quality of common beans. The presence or absence of bean hulls did
not significantly affect protein utilization from tempeh. The PER
for white bean trials (1.47) improved when soaking water was
discarded before the beans were cooked (1.70) and fermented (38).

Conclusions

Common bean protein and procyanidin interactions can be hydrophilic
or hydrophobic, depending on the sites on the protein available for
interaction. Thermal processing can denature the protein and change
the type of interaction possible. Once bean protein is denatured,
hydrophobic interactions between the protein and procyanidin are
likely. Since the strength of hydrophobic interactions increases
with increased in temperature, the interaction between protein and
procyanidin will be enhanced during thermal processing. Removal of
procyanidin will be easiest prior to thermal processing.
     Acute long term doses of procyanidins and food have a reduced
toxicity compared to procyanidin intubated alone. Dietary long term
doses of procyanidins are normally encountered in human dietary
patterns in various areas of the world. Recommendations to increase
common bean consumption will not result in any adverse effects to
populations consuming large quantities of beans.
     Tempeh can be successfully fermented with common beans and
bean/corn mixtures. However, the protein digestibility or
nutritional quality of beans is not improved substantially by tempeh
fermentation.

Acknowledgments

The authors express thanks for assistance from Dr. Kieth Dunker,
Chemistry Department, Dr. Ann Hargis, Veterinary Microbiology and
Pathology and Dr. Robert Bendel, Statistical Services, Washington
State University. Partial financial support for this research was
provided by USAID Title XII Dry bean/Cowpea CRSP. Project No. 0560,
Agricultural Research Center, College of Agriculture and Home
Economics, Washington State University, Pullman, WA 99164-6330.

Literature Cited

1.   Sathe, S. K.; Deshpande, S. S.; Salunkhe, D. K.   CRC Critical
     Reviews Food Sci. Nutr. 1985, 20, 1-46.
2.   Hughes, J. S.; Swanson, B. G.   Food Microstructure 1985, 4,
     183-9.
3.   McLeester, R. C.; Hall, T. C.; Sun, S. M.; Bliss, F. A.
     Phytochem. 1973, 12, 85-93.
4.   Ma, Y.; Bliss, F. A.   Crop Sci. 1978, 17, 431-7.
5.   Buchbinder, B. U.   Ph.D. Thesis, University Wisconsin, Madison,
     1980.
6.   Hosfield, G. L.; Uebersax, M. A.; Isleib, T. G.   J. Amer. Soc.
     Hort. Sci. 1984, 109, 182-9.
7.   Brown, J. W. S.; Osborn, T. C.; Bliss, F. H.; Hall, T. C.
     Theor. Appl. Genet. 1981, 60, 245-250.
8.   Stockman, D. R.; Hall, T. C.; Ryan, D. S.   Plant Physiol. 1976,
     58, 272-5.
9.   Sun, S. M.; McLeester, F. A.; Bliss, F. A.; Hall, T. C.   J.
     Biol.Chem. 1974, 249, 2119-21.
10. Sun, S. M.; Hall, T. C.   J. Agr. Food Chem. 1975, 23, 184-9.
11. Swain, T.; Hillis, W. E.   J. Sci. Food Agric. 1959, 10, 63-8.
12. Price, M. L.; Butler, L. G.   Purdue University Agric. Exp. Sta.
     Bull. No. 272, 1980, p. 37.

13. Reddy, N. R.; Pierson, M. D.; Sathe, S. K.; Salunkhe, D. K.   J. Amer. Oil Chem. Soc. 1985, 62, 541-9.
14. Aw, T-L.; Swanson, B. G.   J. Food Sci. 1985, 50, 67-71.
15. Czochanska, Z.; Foo, L. Y.; Newman, R. H.; Porter, L. J.; Thomas, W. A.; Jones, W. T.   J. C. S. Chem. Comm. 1979, 375-7.
16. Creasy, L. L.; Swain, T.   Nature 1965, 208, 151-3.
17. Haslam, E.   In "Flavonoids and bioflavonoids-current research trends"; Farkas, L.; Gabor, M.; Kallay, F., Eds.; Elsevier Scientific Publishing Co.: N.Y., 1977, pp. 97-110.
18. Brandon, M. J.; Foo, L. Y.; Porter, L. J.; Meredith, P.   Phytochem. 1982, 21, 1953-7.
19. Eastmond, R.   J. Inst. Brewing 1974, 80, 188-92.
20. Lea, A. G. H.   J. Sci. Food Agric. 1978, 29, 471-7.
21. Strumeyer, D. H.; Malin, M. J.   J. Agric. Food Chem. 1975, 23, 909-14.
22. Cansfield, P. E.; Marquardt, R. R.; Campbell, L. D.   J. Sci. Food Agric. 1980, 31, 802-12.

23. Sophianopoulos, J. A.; Durham, S. J.; Sophianopoulos, A. J.; Ragsdale, H. L.; Cropper, W. P.   Arch. Biochem. Biophys. 1978, 187, 132-7.
24. Chamness, G. C.; McGuire, W. L.   Steroids 1975, 26, 538-42.
25. Norby, J. G.; Ottolenghi, P.; Jensen, J.   Anal. Biochem. 1980, 102, 318-20.
26. Tinoco, I.; Sauer, K.; Wang, J. C.   "Physical Chemistry. Principles and applications in biological sciences"; Prentice-Hall, Inc.: Englewood Cliffs, NJ, 1978.
27. Kato, A.; Nakai, S.   Biochim. Biophys. Acta 1980, 624, 13-20.
28. Romero, J.; Ryan, D. S.   J. Agric. Food Chem. 1978, 26, 784-8.
29. Weiland, G. A.; Minneman, K. P.; Molinoff, P. B.   Nature 1979, 281, 114-7.
30. Tanford, C.   "The hydrophobic effect: formation of micelles and microbiological membranes"; John Wiley and Sons: NY, 1980.
31. Bender, A. E.; Mohammidiha, H.   Proc. Nutr. Soc. 1981, 40, 66A.
32. Fairweather-Tait, S. J.; Gee, J. M.; Johnson, I. T.   Brit. J. Nutr. 1983, 49, 303-12.
33. Mitjavila, S.; Lacombe, C.; Carrera, G.; Derache, R.   J. Nutr. 1977, 107, 2113-21.
34. Motilva, M. J.; Martinez, J. A.; Ilundain, A.; Larralde, J.   J. Sci. Food Agric. 1983, 34, 239-46.
35. Rao, B. S. N.; Prabhavathi, T.   J. Sci. Food Agric. 1982, 33, 89-96.
36. Sendzicki, B. J.   M.S. Thesis, Washington State University, Pullman, WA, 1985.
37. Rasyid, A.   M.S. Thesis, Washington State University, Pullman, WA, 1983.
38. Birch, R. E.; Swanson, B. G.; Koos, R. M.; Finney, F.   45th Ann. Mtg. Inst. Food Technol., 1985, Abs. 170.

RECEIVED February 3, 1986

12

Acceptability and Tolerance of a Corn-Glandless Cottonseed Blended Food by Haitian Children

R. E. Hayes[1], Carolyn P. Hannay[2,4], J. I. Wadsworth[3], and J. J. Spadaro[3,5]

[1]Olivet Nazarene College, Kankakee, IL 60901
[2]Grace Children's Hospital, International Child Care, Port-au-Prince, Haiti
[3]Southern Regional Research Center, Agricultural Research Service, U.S. Department of Agriculture, New Orleans, LA 70179

A lysine-fortified, corn-based Public Law 480-type food blend containing glandless cottonseed flour (CC) was found to be comparable in nutritional quality, maternal and child acceptability and child gastrointestinal tolerance to the extensively used U.S. Food-for-Peace Program food blend, corn-soy-milk (CSM). The double blind, four week supplementary feeding study was conducted among 157, mainly preschool age, children and their mothers at nutrition centers in the area of Port-au-Prince, Haiti. The proportion of components in CC were determined by computer to formulate the blend of highest protein quality as measured by chemical score. Animal protein was not used in CC in achieving PER and NPR values statistically comparable to those for CSM.

The feasibility of using cottonseed flour to replace soy flour as a high protein contributor to U.S. Government food blends has been of interest in recent years. Presently cottonseed flour is not used to an appreciable extent in human food in the United States. But, due to expanding population in developing countries, it may be necessary to use this protein resource more efficiently.

An investigation has been conducted to determine the suitability of utilizing glandless cottonseed flour in Public Law 480 (known also as the U.S. Food-for-Peace Program) blended foods. The first stage of this research was to establish a rapid computer formulation procedure to supplant a lengthy trial-and-error approach for determining ingredient proportions needed to achieve best protein quality in the blends (1,2). A series of nine corn-based blends, containing flours of peanut, glandless cottonseed, or cottonseed/soy combinations were then formulated by the computer optimization technique described. Experimental screening tests compared these nine blends with two standard Public

[4]Current address: St. Thomas Hospital, Nashville, TN 37208.
[5]Retired.

Law 480 blends corn-soy and corn-soy-milk, with respect to protein quality by animal assay, organoleptic quality and storage stability. Of the nine experimental blends, a blend of corn-glandless cottonseed, fortified with lysine monohydrochloride, showed a high protein quality and was comparable to corn-soy-milk with respect to overall flavor quality and in degree of flavor maintenance during storage (3).

This present study describes a comparative acceptability and tolerance field test (4), the next step in determining the suitability of using a supplementary food mixture for small children in developing countries. It was conducted in Haiti, mainly among preschool age children. Modified corn-soy-milk (designated MCSM), a sweetened version of the leading U.S. Food-for-Peace Program blended food, was tested against a sweetened experimental blend, corn-glandless cottonseed, fortified with lysine monohydrochloride (designated CC). Haiti was a particularly suitable location for comparative evaluation of these corn-based blends because corn is a major staple there, and because of the availability of nutrition centers that serve mothers with young children. Supplementary foods, such as those provided under Public Law 480, are often distributed at the centers.

Experimental

Blend Formulation,Preparation. Several sources were used to derive the nutritional criteria for prescribing blend compositions and test quantities. Collectively, Protein Advisory Group (designated PAG) guideline numbers 7 and 8 (4,5) and the general U.S.D.A. guidelines for gruel-type foods (6,7) recommend: daily dry weight of supplement; protein concentration and quality; minimum level of fat (for adequate caloric density); maximum levels for crude fiber and total ash; moisture range; and fortification with vitamins, minerals and antioxidants. Because prior experience in field-testing blended foods has shown that addition of sugar improved acceptability, an 8 percent sucrose level was used for both MCSM and CC (8 - 10).

Table I. Specifications for the corn-glandless
cottonseed (CC) and modified corn-soy-milk (MCSM) blends

Component specification	Blend	
	CC (%)	MCSM (%)
Total Protein (all sources)	20.0	20.0
Total Fat (all sources)	6.6	6.6
Sugar	8.0	8.0
Non-fat dry milk	0.0	15.0
Mineral premix	2.7	2.7
Vitamin premix	0.1	0.1

Utilizing proximate analysis data on all blend constituents and amino acid analyses on the protein-containing components, the composition of the corn-glandless cottonseed, lysine monohydrochloide blend was optimized by computer to obtain the best chemical score consistent with the criteria derived from the sources described above and listed in Table I.

The addition of sugar required a different formulation for MCSM from the proportions stipulated by the commodity specification (11). Also, in lieu of the usual procedure of mixing commodities in a given proportion to formulate corn-soy-milk, protein and fat percentage levels of MCSM were set identically to those specified for CC. Cornmeal, defatted soy flour and soy oil proportions were then adjusted by computer to meet these constraints. The vitamin premix provided the antioxidants B.H.A. and B.H.T., each at a level of 0.0022 percent (11) in both blends.

Standard analytical procedures were used to evaluate the composition of ingredients. Of the proximate analyses, nitrogen, lipids, and crude fiber were measured by American Oil Chemists Society (AOCS) methods (12) and moisture and ash by Association of Official Analytical Chemists (AOAC) methods (13). Amino acid analyses were performed by gas-liquid chromatography (14) except for tryptophan, which was analyzed colormetrically (15). In addition to these assays, certain tests of ingredient safety or spoilage were also performed, which space does not permit to be reported in this paper, to assure that ingredients met accepted standards for food safety (16).

The ingredients of each blend were thoroughly mixed in a large ribbon blender. The entire food preparation and packaging operation was carried out in a sanitary manner. A quantity of each blend was appropriately subpackaged for the various physical, chemical and microbiological tests.

In accordance with the recommendation of PAG guideline number 8 (5), the CC and MCSM for field testing were packaged in 0.8 kilogram quantities (approximately 100 grams dry weight of supplement per day for one week). Each 0.8 kilogram batch was weighed into a 3.8 liter (four quart) polyethylene freezer bag, closed with a tie tape, and this bag was then placed within a 1.9 liter (half-gallon) rigid cylindrical polyethylene container with a lid that was sealed with plastic tape. A 53 cc plastic measuring cup, for use in preparing gruel, was included with the food. Numbers were assigned to the CC and MCSM samples on a random basis. There were four 0.8 kilogram packages prepared for each number, one for distribution to each child's mother weekly for four weeks. The plastic containers were packaged for shipment in fiber board drums and held at -18°C until food safety clearances of the U.S. and Haitian governments were obtained. This clearance took approximately nine months.

Chemical,Biological and Physical Tests on Blends. Reported chemical evaluations on the freshly prepared blends include proximate analyses and amino acid analyses. The same analytical procedures described above for ingredients were also used on the blends.

Three animal procedures (protein efficiency ratio, net protein ratio, and protein digestibility) were used to evaluate protein quality. The AOAC (13) animal assay for protein efficiency ratio

(PER) was modified in several respects. The diets were calculated on a 10 percent protein level rather than on an isonitrogenous basis. This was done because the nitrogen factors of the various blend components varied appreciably from the 6.25 nitrogen factor assumed in the AOAC procedure. A composite nitrogen factor for each blend was calculated from analytical results by dividing the total amino acid content by the nitrogen content. In this manner, the composite nitrogen factors were determined to be 6.28 for MCSM and 5.91 for CC. A further deviation from the AOAC PER procedure was that five animals rather than ten were used. The testing laboratory performing the assays had been routinely using five animals for some time for this test and had found only a small difference in the standard error between results for five versus ten animals. The conventional nitrogen factor of 6.25, as specified in the AOAC procedure, was used for computing the Animal Nutrition Research Council (ANRC) casein level.

In the net protein ratio (NPR) calculation (17), 15-day growth and protein intake data of animals on the PER diets were used. Nitrogen digestibility (percent of nitrogen intake absorbed) was determined on each animal on pooled data from the 8th through the 15th day of the PER test.

In addition to the tests described above, other chemical, biological, physical, and physicochemical tests were performed (16) which are not reported in this paper because of space limitation. These tests are mainly concerned with product safety, chemical stability and sensory perception.

Field Test Protocol. Background information on test methodology, nutritional characteristics, and food safety analytical values were required by both the Haitian and American Governments before sending the blended foods to Haiti. In the Haitian Government, approval was required by the Ministry of Health and by its Bureau of Nutrition. Approval of the U.S. Government was given through the U.S. Department of Agriculture's Human Studies Review Committee.

Nutritional criteria for prescribing the blend compositions and quantities used have been described above in the formulation section. Procedural criteria used in the field test are embodied in PAG guideline number 7 for human testing of supplementary food mixtures (4). This guideline requires that human testing be preceded by compositional and nutritional studies of the mixture being considered, assessment of food safety aspects and economic feasibility evaluation. Economic feasibility was not considered in this case since the investigators were interested in a future potential use of glandless cottonseed, which is not now economically competitive with soy. Compositional, nutritional, and food safety studies were previously conducted as described above (3). The same types of evaluation were performed on foods actually sent to Haiti. The Haitian field trial concerns one of the four categories of human evaluations outlined by PAG guideline number 7 (4), namely, acceptability and tolerance tests. Some of the features of these tests are that: at least some of the testing should take place in a country for which the protein-rich food is intended; the test should be conducted with children in the age categories for which the product is intended; both the test and control groups should be of similar size; at least 20 individuals should be used in the test; the duration of the test should be at least 4 weeks; data analysis should consider that the mother's acceptability response may influence the child's response; and the possibility that disease processes might influence acceptability and tolerance should be considered. Refusal of the child to eat the food is considered to be an indicator of poor palatability. Intolerance is judged by noting persistent gastrointestinal upsets. It is also suggested that other clinical responses, such as allergic reactions, be recorded.

Consideration of the provisions of PAG guideline number 7 determined test procedure, to be described, and also the composition of the questionnaire. The questionnaire actually used was a Creole translation of the one shown in Figure 1.

The questions were to be asked of each mother for each child by the nutrition clinic worker after the four-week feeding trial. Question A concerns the child's acceptability response. The child's eagerness to eat the food is used as a gauge of how well the child liked the food. Item B reflected the mother's overall acceptability response to the food. The mother's evaluation of four factors influencing her acceptability is provided for in item C. The factors under item D involve the mother's estimate of whether certain indicators of gastrointestinal reaction change and the direction of the change during the interval the blended food was consumed. The mother was also asked (item E) to provide information that would help investigators to check possible relationships between questionnaire responses and other foods eaten during the test interval.

The blended foods were distributed to mothers at five nutrition centers in the Port-au-Prince area of Haiti. One of the coauthors, Carolyn P. Hannay, R.N., nutritionist, supervised food distribution and collection of data.

All children were to be 5 years of age or younger. Mothers had from one to four children participating in the field test. The mothers were grouped according to the number of children that they had participating in the study. A randomization procedure was used for sample assignment to mothers with two children in the study, and then for mothers with three children, et cetera. For each mother designated to receive a given blend (eg., MCSM), each child in her family in the test received different sample numbers of the same blend. Different blends were not assigned to the same mother.

Mothers were given a demonstration on how to prepare gruel from a dry blend. Written directions in Creole were made available for use by nutrition workers. The feeding was to be done three times per day. Some flexibility was allowed the mother in preparation of the gruel. The same reconstitution directions were given to mothers with either MCSM or CC, even though it was known that consistencies of the two blends differed. No clue was given that there were two blends.

The experiment was double-blind. The auxiliary nutritionist distributing the food containers to the mothers did not know which blend a certain number represented. She only knew that a given mother received a sample with a particular assigned number for a certain child. Nor, of course, was the mother told that the blended food she received was of any particular type. A 0.8

BLENDED FOOD QUESTIONNAIRE

Questions to be asked of each mother for each child by the nutrition clinic worker after the four-week feeding trial.

Family name:_____

Child's first name:_____Age_____

How many children of this family are participating in this feeding study? _____

Blended food sample number:_____

A. How did your child like the blended food as indicated by eagerness to eat the food?

Like	O.K.	Dislike

Put an "X" in one block

Any comment(s)?_____

B. How did you like the blended food?_____

Like	O.K.	Dislike

Put an "X" in one block

C. Did you consider the blended foods:

	Good?	Fair?	Poor?
Appearance			
Flavour			
Feel-in-the-mouth			
Ease of preparation			

D. During the four weeks the new food was eaten, did:

	Increase?	Decrease?	Remain the same?
Appetite			
Flatulence			
Vomiting			
Diarrhea			
Undigested stool contents			

E. What other foods did your child consume during the four weeks the food supplement was eaten?

_____
_____
_____

Figure 1.   Blended food questionnaire.   (Reproduced with permission from reference 16.   Copyright 1983 United Nations U.)

kilogram quantity of dry blend, of the same sample number, was provided weekly for each of four weeks for each child. At the end of the four weeks, the nutrition worker at each clinic, using the Creole translation of the questionnaire described, recorded the response of each mother, and that of her child, to the assigned blended food sample. The auxiliary nutritionist also recorded other foods consumed during the test interval, because the blended foods were meant to supplement the regular diet.

Statistical Analysis. For each of the 11 possible questionnaire responses for preference or gastrointestinal effect, as shown in Figure 1, value numbers were assigned to the different descriptive responses. For the section on the childs eagerness to eat and mother's overall liking of the food, values of 1, 2 and 3 were assigned to the responses Like, OK and Dislike, respectively. For the section on mother's opinion of food characteristic, values of 1, 2 and 3 were assigned to the responses Good, Fair and Poor, respectively. For the section on mother's observation of gastrointestinal effects, values of 1, 2 and 3 were assigned to the responses Increase, Remains the same, and Decrease, respectively. In addition, for the purpose of statistical analysis, the nutrition center Christ-Roi (69 Children) was designated No.1; Carrefour Feuilles (34 children) was No. 2; and the combined Delmas, Pernier and Salvation Army centers (54 Children) was No. 3.

Several approaches were taken to handling the data by analysis of variance (ANOVA) (18). In some cases, data from all participating children were used in the analysis. In other instances a hierarchal design was employed that used only part of the collected data.

Combining the data from all the clinics, ANOVA was used to test the 11 questionnaire responses for statistical significance of variations among the three nutrition centers, and between the two blend types, and whether a difference between blends depended upon the nutrition center (interaction effect). The same variations were examined statistically using only the first child in each family. The rationale for this approach was that data from different children of the same family might be correlated, because the same mother responded to all the questions.

Several circumstances developed during the field test that necessitated repeating the ANOVA after certain data were omitted. For a series of samples, more than one child used the same sample number. This problem is understandable in a setting with a high incidence of protein-calorie malnutrition (19). There were also a few children in the study who were above the originally set upper age limit of five years. ANOVAs were made excluding both observations on children sharing the same numbered sample and those on children over five years. In making these reruns, appropriate adjustments had to be made in some retained data because of changes in child-order within family and total number of children in the family participating in the study.

Results and Discussion

Composition and Nutritional Quality. Comparisons of blends CC and MCSM are summarized in Tables II through IV. Table II shows that

there were some marked differences between blends in protein providing ingredients, but added oil and fortification levels,
except for lysine monohydrochloride were similar. The compositions

Table II. Formulation of the field tested food blends
corn-cottonseed (CC) and modified corn-soy-milk (MCSM).

Ingredient	Amount in Blend (%)	
	CC	MCSM
Cornmeal, processed, gelatinized	52.5	45.4
Soy flour, defatted, toasted	--	22.9
Cottonseed flour, glandless, defatted	31.5	--
Milk, nonfat dry	--	15.0
Sucrose, granulated	8.0	8.0
Soy oil	--	5.9
Cottonseed oil	5.1	--
Mineral premix	2.7	2.7
Vitamin premix with antioxidants	0.1	0.1
L-lysine HCl	0.09	--

of the mineral premix and vitamin premix with antioxidants are
given in reference (3). The proximate analyses (Table III) are
very similar for both blends. The ingredient percentages and
similar proximate analyses of both blends aligned quite well with
the specifications previously described. The protein quality and
caloric densities of the two blends are given in Table IV.
Although there was a difference in the most limiting amino acids
between the blends, they were close in chemical score. The
chemical score is the score of the most limiting essential amino

Table III. Proximate analyses of field tested
food blends corn-cottonseed (CC) and modified
corn-soy-milk (MCSM).

Component	Amount in Blend (%)	
	CC	MCSM
Protein	20.7	20.8
Lipids	6.3	6.5
Crude fiber	0.9	1.0
Ash	4.8	5.3
Moisture	8.7	8.8
Carbohydrate (by difference)	58.6	57.6

acid. The amino acid scores were computed according to the
definition in reference (20). The two blends were not found to be
statistically significantly different in PER, proportionally
adjusted to reference casein, or in NPR. The blends were almost
equal in caloric density, an important characteristic of weaning
foods. Overall, the nutritional values of both blends were very
comparable ·

Table IV. Nutritional quality of field tested food blends
corn-cottonseed (cc) and modified corn-soy-milk (MCSM).

Quality factor	Blend	
	cc	MCSM
Chemical score	80	84
Limiting amino acid (s)	Threonine	Methionine +cystine
PER(adjusted) + S.E.	2.12+0.05	2.26+0.06
NPR + S.E.	3.62+0.07	3.72+0.05
Nitrogen digestibility (%)	87.7	85.6
Caloric density (Kcal/100g)	374	372

Acceptability and Tolerance Responses. The results of statistical
analyses of the blended food questionnaire responses are presented
in Table V. In this table, testing by analysis of variance of
blend type difference is designated by BT, of clinic number
difference by CN, and of the dependence of differences between
blends on clinic number by CN-BT.

Reference is now made to the first vertical set of results in
Table V, encompassing all clinics and all children. One hundred
fifty seven children were involved; 77 used blend CC and 80, blend
MCSM. It is apparent, the mean scores nearly equal to one, that
both blends were very acceptable to both children and mothers.
Despite the fact that blend CC had a slightly green cast due to its
cottonseed flour component, color difference between blends did not
affect the mothers' comparative appearance acceptability.
Appearance and flavor of both blends received the highest possible
acceptability rating by the mothers. They had a significantly
higher preference (p<0.05) for the mouth feel of the MCSM blend,
which was thinner in consistency than the CC blend. Despite this
statistically significant difference, the mean rating values for
blend MCSM (1.013) and for blend CC (1.132) were close, and both
were near the best rating score scale of 1 = good. For both
blends, appetite increased during the feeding trial. The mean
values showed that the incidence of flatulence, vomiting, diarrhea
and the amount of undigested stool contents remained about the same
during the interval when the food supplements were consumed.

Using the same data base of all children from all clinics,
there were no statistically significant interaction effects found,
wherein differences between blends depended on the nutrition
center. Statistical significance was also tested for among the
variations in ratings of the nutrition centers. Although, as
previously shown, there were no statistically significant
differences found between blends for the mothers's evaluation of
appetite and flatulence changes, there were highly significant
differences (P<0.01) in rating values for these characteristics
among clinics, as shown in Table V. Mothers at clinic number 2, for
some reason, less often noted an increase in appetite when children

Table V. Results from blend food questionnaire for corn-cottonseed (CC) and modified corn-soy-milk (MCSM). Statistical significance tested by blend type (BT), by clinic (CN), and by interaction of blends on clinic (CN-BT).

Response factor	All clinics combined, all children					All clinics, unqualified children omitted				
	Blend mean score		Statistical significance			Blend mean score		Statistical significance		
	CC	MCSM	BT	CN	CN-BT	CC	MCSM	BT	CN	CN-BT
Child's eagerness to eat	1.091	1.088	NS	NS	NS	1.059	1.088	NS	NS	NS
Mother's overall liking	1.195	1.100	NS	NS	NS	1.216	1.088	NS	NS	NS
Mother's opinion:										
Appearance	1.000	1.000	NS	NS	NS	1.000	1.000	NS	NS	NS
Flavor	1.000	1.000	NS	NS	NS	1.000	1.000	NS	NS	NS
Feel-in-the-mouth	1.132	1.013	*	NS	NS	1.020	1.018	NS	NS	NS
Ease of preparation	1.013	1.013	NS	NS	NS	1.000	1.000	NS	NS	NS
Gastrointestinal effects:										
Appetite	1.273	1.269	NS	**	NS	1.294	1.236	NS	**	NS
Flatulence	1.957	1.947	NS	**	NS	1.933	1.923	NS	**	NS
Vomiting	1.986	2.013	NS	NS	NS	2.000	2.019	NS	NS	NS
Diarrhea	1.972	1.960	NS	NS	NS	1.935	1.923	NS	NS	NS
Undigested stool	1.942	1.880	NS	NS	NS	2.000	1.904	NS	*	NS

By clinic, all children:

Response factor	Clinic number	Blend mean score		Stat. sig.
		CC	MCSM	BT
Appetite	1	1.210	1.314	NS
	2	1.722	1.667	NS
	3	1.154	1.000	NS
Flatulence	1	2.000	2.000	NS
	2	1.833	1.800	NS
	3	2.000	1.960	NS

*--significant at p < 0.05
**--significant at p ≤ 0.01
NS--not significant

were consuming either blend. Mothers at clinic number 2 also more often noted increased flatulence when either blend was consumed. However, in the case of both appetite and flatulence, significant differences between the two blends at clinic number 2 could not be found.

Table V also decribes results of analyses of variance that were performed when data for unqualified children were removed. The data for 42 children were excluded from the total of 157, because of sharing food having the same sample number. The data for 7 children, age 6 and older, were also excluded. When data in both these categories were excluded, significant differences were not found between the two blends for any of the 11 questionnaire responses, including feel-in-the-mouth. Perhaps, because of fewer observations, a significant difference between blends was not found for this characteristic when unqualified children were excluded from the total. As in the case when all children's responses were included in the analysis, statistically highly significant (<0.01) differences in ratings among clinics were found for appetite and flatulence. In addition to this, there was a statistically significant (p<0.05) difference in score among clinics (produced by responses in clinic number 1) for undigested stool contents. But again, there was no significant difference between blends for this characteristic.

In addition to Table V reported results, analyses of variance were performed on data from all clinics, using only the first child in each family. In this case also, statistical significance between blends was not found for any of the 11 response characteristics.

The assigned blended food was intended as a supplement to the regular diet of each child during the feeding trial. The survey of mothers at the termination of the field test showed that the most popular cereal staple was corn, followed by rice. Beans or other type vegetables were also commonly eaten. The consumption of beans by the children during the trial might have masked any effect the cottonseed flour had on flatulence. In comparison to plant foods, animal protein foods such as milk, meat, and fish were consumed by a considerably lower percentage of the children. No allergic responses were observed during the study.

Summary

This study has demonstrated the feasibility of producing a corn-based Public Law 480-type food blend containing glandless cottonseed flour that is comparable in nutritional quality, maternal and child acceptability, and child gastrointestinal tolerance to the extensively used U.S. Food-for-Peace Program food, corn-soy-milk.

The blend containing glandless cottonseed flour offers the economic advantage of not requiring an animal protein component. However, glandless cottonseed flour is not at present time economically competitive with soy flour. With more extensive cultivation of glandless cottonseed, the prospect of using this commodity in nutritious food blends will be more favorable.

Literature Cited

1. Hayes, R.E.; Wadsworth, J.I.; Spadaro, J.J. Cereal Foods
    World 1978, 23, 548-556.
2. Wadsworth, J.I.; Hayes, R.E.; Spadaro, J.J. Cereal Foods
    World 1979, 24, 274-286.
3. Hayes, R.E; Spadaro, J.J.; Wadsworth, J.I.; Freeman, D.W. "A
    study of nine Corn-Based Food Blends Formulated by Computer
    for Maximum Protein Quality," Agri. Research Results
    (South.Ser.) ARR-S-18, U.S. Department of Agriculture, 1984.
4. FAO/WHO/UNICEF, Human Testing of Supplementary Food
    Mixtures," PAG guideline no. 7, 1970.
5. FAO/WHO/UNICEF, "Protein-Rich Mixtures for use as Weaning
    Foods," PAG guideline no. 8, 1971.
6. Senti, F.R. In "Protein Enriched Cereal Foods for World
    Needs"; Milner, M., Ed.; American Association of Cereal
    Chemists: St. Paul, MN, 1969; pp. 246-254.
7. Senti, F.R. Cereal Science Today 1972, 17, 157-159, 161.
8. Combs, G.F. PAG Bull. 1967, 7, 15-24.
9. Abrahamson, L.; Hambraeus, L.; Valquist, B. PAG Bull. 1974,
    4, 26-30.
10. U.S. Agricultural Stabilization and Conservation Service
    "Sweetened, Instant Corn-Soya-Milk Announcement CSM-6 (with
    updating through amendment 7)," 1982.
11. U.S Agricultural Stabilization and Conservation Service,
    "Corn-Soya-Milk Export Announcement CSSM-1 (with updating
    through amendment 5),"1982.
12. American Oil Chemists Society (AOCS), "Official and Tentative
    Methods (with updating through 1982)," 1982, 3rd ed.
13. Association of Official Analytical Chemists (AOAC), "Official
    Methods of Analysis," 1982, 12th ed.
14. Kaiser, F.E.; Gehrke, C.Z.; Zumwal, R.W.; Kuo, K. J.
    Chromatogr. 1974, 94, 113-133.
15 Amaya-F, J.; Young, C.T.; Chichester, Q.O. J. Agric.
    Food Chem. 1977, 25, 139-142.
16. Hayes, R.E.; Hannay, C.P.; Wadsworth, J.I.; Spadaro, J.J.
    Food and Nutr. Bull. 1983, 5 (3), 23-34.
17. Jansen, G.R. Food Technol. 1978, 32 (12), 52-56.
18. Snedecor, G.W.; Cochrane, W.G. "Statistical Methods"; Iowa
    State University Press: Ames, IA, 1980, 7th ed.; pp. 258-338.
19. Berggren, G.C. Food Nutr. Bull. 1981, 3 (4), 29-33.
20. Joint FAO/WHO Ad Hoc Expert committee on Energy and Protein
    Requirements. "Energy and Protein Requirements"; WHO Tech.
    Rep. Ser. 522; FAO Nutr. Meeting Rep. Ser. 52, 1973.

RECEIVED January 24, 1986

13

Influence of Animal and Vegetable Protein on Serum Cholesterol, Lipoproteins, and Experimental Atherosclerosis

David M. Klurfeld and David Kritchevsky

The Wistar Institute of Anatomy and Biology, Philadelphia, PA 19104

One distinguishing characteristic between protein of
animal and vegetable origin is the ratio of lysine to
arginine (L/A). Feeding casein (L/A-1.9) to rabbits
gave cholesterol levels of 268 mg/dl and 1.34 average
score of aortic atherosclerotic lesions (0-4 scale); soy
isolate (L/A-0.9) gave 145 mg/dl cholesterol and 0.55
lesions. Addition of lysine to soy to give L/A ratio of
casein raised lipids and lesions significantly. Addition
of arginine to casein had a slight lowering effect on
lipids and atheroma. Feeding three animal proteins of
varying L/A ratios (fish, 1.4; casein, 1.9; milk, 2.4)
resulted in atherosclerosis significantly correlated (r =
0.99) with L/A ratio. Feeding beef or textured vegetable
protein (TVP) yielded significantly higher serum cholesterol and atherosclerosis in the group fed beef; however, an equal mixture of beef:TVP gave results similar
to TVP alone. Animal proteins are more atherogenic
than those from vegetable sources and the L/A ratio may
give an index of magnitude of effect. When animal protein is part of a mixed diet, the atherogenic effect is
diminished or abrogated.

The first experimental studies on nutritional induction of
atherosclerosis were performed by Ignatowski (1) who fed rabbits
milk, beef, and egg yolk. These experiments were interpreted as
indicating that the animal foods consumed by the wealthy contributed
to atherosclerotic heart disease which was observed only among the
upper class at that time. However, Anitschkow and Chalatow (2)
published studies shortly thereafter in which they fed purified
cholesterol to rabbits and achieved the same results which led them
to conclude that the sterol present in foods of animal origin, was
the predominant factor in genesis of the arterial lesions. This
finding has had lasting and significant influence on the direction
of research about atherogenesis. The earliest comparison of animal
and plant proteins in a model of experimental atherogenesis showed
that feeding casein led to more atherosclerosis than feeding soy

0097-6156/86/0312-0150$06.00/0
© 1986 American Chemical Society

protein whether or not cholesterol was added to the diets (3).
Since the publication of this experiment over 45 years ago, there
has been a recent resurgence of interest in the area of protein
effects on plasma lipids and atherosclerosis. This paper summarizes
some of the data generated in our laboratory over the last several
years and discusses their significance in relation to
hypercholesterolemia and atherosclerosis in other species including
humans.

The standard diet used in our experiments is a semipurified,
cholesterol-free preparation that is composed of 25% protein, 40%
sucrose, 13% coconut oil, 1% corn oil, 15% cellulose, 5% mineral
mix, and 1% vitamin mix. This diet has been shown to induce an
endogenous hypercholesterolemia and lead to atherosclerosis in
rabbits and monkeys (4,5). The specific question addressed by our
series of investigations is whether the type of dietary protein,
when all other dietary components are constant, can influence the
development of hyperlipoproteinemia and atherosclerosis. More
specifically, we have examined the effects of the individual amino
acids, lysine and arginine, and their ratios in the diet on plasma
and hepatic lipids as well as the development of arterial plaques.
To study amino acid effects, we have added single amino acids to the
diets to approximate the lysine/arginine (L/A) ratio of different
proteins. For example, the L/A ratio of casein is about 2.0;
therefore, enough lysine was added to the diet containing soy
protein isolate to bring its endogenous L/A ratio of 0.9 up to that
of casein. Similar manipulations were performed with addition of
arginine to casein to give the L/A ratio normally found in soy
protein.

Four experimental diets based upon that described above were
fed to rabbits for eight to ten months in three separate
experiments. The results are summarized in Table I. Serum
cholesterol concentrations in rabbits fed casein were significantly
elevated over those observed with soy protein feeding. Addition of
arginine to casein had inconsistent effects on serum cholesterol
levels; addition of lysine to soy protein usually raised the
cholesterol levels above those found by feeding soy alone. Serum
triglycerides were significantly lower in the animals fed soy than
in the rabbits fed casein. Addition of the individual amino acids
to the proteins had no consistent significant effects on
triglycerides. The data for aortic atherosclerosis are expressed on
a 0 to 4 scale with 4 denoting complete coverage of the intimal
surface with lipid-containing lesions. Table I shows that feeding
of soy protein leads to less than half the aortic atherosclerosis
found when casein is the protein. Addition of arginine to casein
reduces the severity of aortic atheroslerosis by about 25% while
addition of lysine to soy protein increased the severity of
atherosclerosis in the aortic arch by 57% and that in the thoracic
aorta by 75%.

Serum samples from some animals fed the four dietary regimens
were pooled and subjected to fractionation of the lipoproteins by
agarose column chromatography. The total amounts of lipoproteins
found in the sera from the four groups (ug/ml) were: casein, 904;
soy protein, 807; casein/arginine, 1,130; and, soy/lysine, 672.
Table II gives the distribution of lipoproteins in the various
density classes from these animals. Since the total amounts of

lipoproteins in the sera of rabbits fed the diets containing different proteins was so varied, a comparison of the percentage distribution of the lipoprotein classes is more appropriate than simply comparing absolute amounts. The percentages of very low density (VLDL) and intermediate density lipoproteins (IDL) were similar in the animals fed the casein and soy/lysine diets and were double the percentage of lipoprotein in these two density classes found in the sera of animals fed soy protein or casein/arginine diets. The percentages of low density lipoproteins (LDL) were similar in all groups and the percentage of high density lipoprotein (HDL) was highest in the animals fed soy protein.

Table I. Serum Lipids and Aortic Atherosclerosis in Rabbits fed
Casein, Soy Protein, Casein & Arginine, or Soy & Lysine

Treatment	Number of Rabbits	Serum Lipids (mg/dl)		Atherosclerosis	
		Cholesterol	Triglycerides	Arch	Thoracic
Experiment 1					
Casein	7	174 + 30	133 + 17	2.2 + 0.5	1.5 + 0.4
Soy protein	7	77 + 21	98 + 17	1.1 + 0.4	0.7 + 0.3
Cas/arginine	6	129 + 12	186 + 20	1.4 + 0.4	0.8 + 0.3
Soy/lysine	6	106 + 29	101 + 14	1.6 + 0.4	1.1 + 0.2
Experiment 2					
Casein	8	283 + 28	81 + 11	1.1 + 0.3	0.8 + 0.3
Soy protein	11	234 + 20	53 + 7	0.5 + 0.2	0.4 + 0.1
Cas/arginine	7	343 + 65	117 + 58	1.3 + 0.4	0.7 + 0.2
Soy/lysine	11	242 + 22	70 + 7	0.9 + 0.3	0.5 + 0.2
Experiment 3					
Casein	5	377 + 59	104 + 28	1.6 + 0.2	0.9 + 0.3
Soy protein	7	117 + 10	43 + 4	0.6 + 0.2	0.1 + 0.1
Cas/arginine	7	271 + 74	75 + 13	1.2 + 0.2	1.3 + 0.2
Soy/lysine	8	242 + 43	60 + 5	0.5 + 0.2	0.7 + 0.1

Table II. Percentage Distribution of Lipoproteins from Rabbits
fed Various Proteins

Treatment	VLDL	IDL	LDL	HDL
Casein	2.3	14.4	31.9	51.4
Soy protein	1.1	7.7	30.0	61.2
Cas/arginine	0.9	7.8	35.8	55.5
Soy/lysine	2.2	17.8	30.2	49.9

Another experiment compared the effects of feeding partially defatted beef with textured vegetable protein (TVP) derived from soy beans. In this experiment, the fat was beef tallow and it was added so that all groups had a total of 14% fat in the diet. This allowed compensation for the fat content of the partially defatted beef so that only 9.3% tallow was added to the beef diet. Similarly, carbohydrate and crude fiber were adjusted in the diets containing TVP so that each group consumed 40% digestible carbohydrate and 15% crude fiber. In addition to the groups fed beef and TVP, a third group was fed an equal mixture of the two protein sources. The diets were fed to groups of twelve rabbits for eight months. At the end of the study, the animals were killed and serum and hepatic lipids as well as aortic atherosclerosis were quantitated. Table III gives some of the data from this experiment.

Table III.   Serum Lipids and Aortic Atherosclerosis in Rabbits fed Beef, TVP, or an Equal Mix of Beef & TVP

	Beef	TVP	Beef-TVP
Serum lipids (mg/dl)			
Cholesterol	185 + 24	37 + 4	61 + 6
% HDL-cholesterol	20 + 2	39 + 4	43 + 4
Triglycerides	60 + 8	59 + 7	70 + 13
Aortic atherosclerosis			
Arch	1.3 + 0.2	0.8 + 0.1	0.7 + 0.1
Thoracic	0.8 + 0.1	0.2 + 0.1	0.4 + 0.1

Statistically significant differences were observed in serum total cholesterol, HDL-cholesterol, and aortic atherosclerosis. Serum triglycerides were unaffected by dietary protein. Textured vegetable protein as the sole dietary protein source resulted in one-fifth the amount of serum cholesterol and about half the atherosclerosis in the animals fed beef. Most importantly, the animals fed an equal mix of beef and TVP had serum cholesterol levels and atherosclerosis almost equal to those seen with pure vegetable protein. HDL-cholesterol concentrations were similar in the all TVP and beef-TVP dietary groups which were double that seen in the all beef group. The results obtained for serum lipids and atherosclerosis with feeding beef were almost identical to a group fed casein concomitantly as the sole protein source (6). In this particular experiment, a small amount of cholesterol was consumed with the beef and tallow. However, due to the design of the diets which made up for the endogenous fat content of the beef, the cholesterol intake (10-12 mg/day) was similar in all groups. Furthermore, at this level of cholesterol intake little or no effect would be expected on plasma lipids and the development of atherosclerosis (3).

An important point relevant to the above results is that the diet was identical for all groups within a given experiment except for protein source. This is necessary to distinguish specific nutrient effects but does not reflect actual food consumption since intake of animal protein is also associated with increased ingestion of total fat, saturated fat, and cholesterol. Therefore, it is only

through manipulation of diets of experimental animals that the
question of protein effects on lipids and atherosclerosis can be
addressed in an adequate manner. It is difficult to extrapolate
findings in rabbits to humans but other investigators have confirmed
our results in several animals species. Studies with humans fed soy
protein have also been reported but these have been less consistent
than the animal studies for reasons that will be discussed below.
     Kim et al (7) fed swine diets containing either casein or soy
protein. These diets resembled the typical American diet in that
they derived about 42% of calories from fat and 21% of calories from
protein with the remainder from carbohydrate. In these test diets,
90% of the fat was butter and the rest was corn oil. In addition,
crystalline cholesterol was added to bring total cholesterol in each
diet to 1055 mg per day. These investigators also fed one group of
swine a diet with equal parts of casein and soy. Feeding these
diets for four or six weeks led to a doubling of serum cholesterol
concentrations when the protein was casein but only a small
nonsignificant rise when soy protein or the mixture of soy and
casein was fed. In addition, these researchers found that
substitution of soy protein for casein completely inhibited the
increase in total body cholesterol that they had demonstrated in
swine fed a casein-based diet. Further studies by this group of
investigators revealed that the hypocholesterolemic effect of soy
protein was evident in swine only when they were fed a
hypercholesterolemic diet (8). Inclusion of soy protein in the
basal hypercholesterolemic diet resulted in significant increases in
fecal excretion of neutral and acidic steroids that was not
accompanied by increases in cholesterol synthesis. The difference
between soy and casein was not observed when included in a basal
mash diet, indicating that this dietary effect in swine was
important only under conditions favorable to net cholesterol
retention.
     Semipurified diets whose basic composition was similar to
diets consumed in Western countries were fed to cynomolgus monkeys;
the differences between the groups were that one diet contained
casein and the other contained soy protein isolate (9). In a
crossover experiment, soy protein was exchanged for casein and vice
versa for three-week periods. Although there were no significant
differences in total serum cholesterol concentrations, beneficial
effects on individual lipoprotein classes were demonstrated
following feeding of the soy protein diet. HDL cholesterol levels
were significantly higher and VLDL cholesterol levels were
significantly lower when soy protein replaced casein.
     Since most studies concerning alterations in lipid metabolism
among animals fed different types of protein have been performed
using rodents, the remainder of this paper will concentrate on those
findings. Carroll and his associates have investigated this area
extensively. Hamilton and Carroll (10) surveyed a variety of animal
and plant proteins for their effects on cholesterol levels in
rabbits fed a cholesterol-free, semipurified diet. These proteins
were delipidized by solvent extraction and fed to the animals for
28-day periods. Although there was relatively wide variation in the
response of serum cholesterol to feeding of the proteins, ingestion
of animal proteins resulted in significantly higher serum
cholesterol concentrations than did consumption of plant proteins

(Table IV). Huff et al (11) extended these observations to include
enzymatic hydrolysates of the native proteins and amino acid
mixtures equivalent to soy or casein. In this experiment amino acid
contents were identical among the diets fed as the three forms of
each protein but the structure was different (Table V).

Table IV. Serum Cholesterol Levels in Rabbits fed Different
Proteins

Protein	Serum Cholesterol (mg/dl)
Animal protein	
Whole egg	235 + 89
Skim milk	230 + 40
Lactalbumin	215 + 69
Beef	160 + 60
Pork	110 + 17
Egg white	105 + 28
Plant protein	
Wheat gluten	80 + 21
Peanut protein	80 + 10
Soy protein concentrate	25 + 5
Promine-R	15 + 5

From: Hamilton and Carroll (10)

Table V. Serum Cholesterol Levels in Rabbits fed Casein, Soy
Protein, Their Hydrolysates and Their Amino Acid Mixtures

Dietary treatment	Serum Cholesterol (mg/dl)
Intact casein	213 + 53
Enzymatic hydrolysate	178 + 30
Amino acid mixture	213 + 42
Intact soy protein	69 + 12
Enzymatic hydrolysate	41 + 8
Amino acid mixture	124 + 30

From: Huff et al (11)

Table VI. Serum Cholesterol Levels in Rabbits fed Mixtures of Soy
and Casein

% Casein / % Soy Protein		Serum Cholesterol (mg/dl)
100	0	239 + 28
75	25	138 + 20
50	50	70 + 15
0	100	67 + 12

From: Huff et al (11)

While the enzymatic hydrolysate of casein and its equivalent amino acid mixture gave serum cholesterol levels similar to intact casein, such was not the case with feeding soy protein. The enzymatic soy hydrolysate was almost equivalent to the intact protein but feeding the amino acid mixture gave serum cholesterol levels double that of the native protein (although still significantly lower than feeding casein). This indicates that the tertiary structure of the protein, or some minor non-protein contaminant, has a significant effect of serum cholesterol levels. The experiment in which dilution of casein with soy protein was studied (Table VI) showed a reduction in serum cholesterol proportional to the amount of soy protein in the diet up to 50% which was equal to an all soy diet. If a contaminating factor were present, it would have to be extraordinarily potent to exert such an effect. When the soy and casein diets were fed for ten months significant differences in aortic atherosclerosis were found (12). Feeding casein resulted in average atheroma scores of 1.8 + 0.3 (0-4 scale) while soy protein gave a mean value of 0.2 + 0.06, which is significant at p < 0.01.

Terpstra and his associates have been active in investigating the effects of dietary protein and single amino acids on serum cholesterol levels in several species (reviewed in 13). They confirmed that feeding casein results in significantly higher serum cholesterol concentrations than does soy protein. In their studies, addition of 0.8% arginine to a 20.8% casein diet raised serum cholesterol of rabbits, in contrast to our findings. Addition of glycine resulted in half the serum cholesterol obtained with feeding casein alone. Interestingly, a diet containing a mixture of three animal proteins (8% casein, 5.6% gelatin, and 8% fish protein) gave serum cholesterol levels slightly higher than those found with feeding soy protein and significantly lower than those observed when casein was fed alone.

A characteristic of casein and other animal proteins that distinguishes them from plant proteins is the high lysine content. We have demonstrated that addition of lysine to soy protein causes a significant increase of serum cholesterol and development of atherosclerosis. Another approach to this question was to feed three proteins of animal origin that contained almost identical amounts of lysine but different amounts of arginine. We used fish protein (lysine, 6.81; arginine, 4.74; L/A ratio, 1.44), casein (lysine, 6.91; arginine, 3.65; L/A ratio, 1.89), and milk protein (lysine, 6.61; arginine, 2.71; L/A ratio, 2.44) (14). The proteins

Table VII. Serum Lipids and Atherosclerosis in Rabbits fed Three Animal Proteins with Different Lysine/Arginine Ratios

	Fish protein	Casein	Milk protein
Cholesterol (mg/dl)	238 + 40	530 + 76	462 + 62
Triglycerides (mg/dl)	122 + 20	177 + 47	251 + 56
Atherosclerosis			
Arch	1.6 + 0.2	2.1 + 0.3	2.6 + 0.2
Thoracic	1.0 + 0.2	1.1 + 0.3	1.6 + 0.2

From: Kritchevsky et al (14)

were fed at 25% in our standard cholesterol-free, semipurified diet
for eight months. Results from this study are given in Table VII.
Analysis of linear regression between aortic atherosclerosis and the
lysine/arginine ratio of the three proteins yielded r = 0.99 (p <
.05). These results indicate that not only does the amount of lysine
affect serum cholesterol and atherosclerosis but in proteins with
similar lysine contents the proportion of lysine to arginine may be
a significant determinant of their effects on these parameters.
Arginine has been reported to be a glucagon secretagogue and may
have glucagon-like activity (15). Sugano et al (15) reported that
addition of arginine to casein resulted in a significant increase in
plasma glucagon levels in rats. In this study, casein and soy
feeding yielded similar levels of glucagon but significantly lower
immunoreactive insulin concentrations were found in the rats fed soy
protein. Since peptides formed during protein digestion can
function as glucagon secretagogues through some mechanism other than
alterations of amino acid levels in the plasma (16), it is necessary
to study differences these proteins may exert in the intestine. To
this end, we have examined the influence of casein and soy protein
on intestinal absorption and lymphatic transport of cholesterol and
oleic acid (17). Although overall recovery of cholesterol and oleic
acid was similar among animals fed different proteins, analysis of
timed lymph collections revealed that absorption of lipids was more
rapid in rats fed casein than in those fed soy protein.
Furthermore, addition of lysine to soy protein markedly increased
the rate of lipid absorption and addition of arginine to casein
slowed lipid absorption. The slowed absorption of lipids in animals
fed soy protein is similar to that reported for soluble fibers such
as pectin and guar gum that act to lower serum cholesterol
concentrations in a number of animal species, including humans.

    In addition to more rapid absorption of lipids in animals fed
casein, another mechanism that may be operative is decreased
clearance of circulating lipids. Rabbits fed a casein-based
semipurified diet excreted significantly less cholesterol but more
bile acids in their feces than animals fed a commercial diet (18).
The total sterol excretion in feces of the animals fed the casein
diet was half that of the rabbits fed the stock diet. Huff and
Carroll (19) found that rabbits fed soy protein had a much faster
turnover rate of cholesterol and a significantly reduced rapidly
exchangeable cholesterol pool compared with rabbits fed casein.
Similar studies performed in our laboratory revealed that the mean
transit time for cholesterol was 18.4 days in rabbits fed soy
protein, 36.8 days in rabbits fed casein, 33.7 days in rabbits fed
soy plus lysine, and 36.3 days in rabbits fed casein plus arginine.
These data suggest that addition of lysine to soy protein

Table VIII. Dietary Protein Effects on Cholesterol Kinetics in
                           Rabbits

	Pool A (mg)	Pool B (mg)	Mean transit time (days)
Soy protein	427	711	18.4
Soy & lysine	791	1260	33.7
Commercial diet	200	690	11.8

significantly inhibited cholesterol flux and this was borne out by
the finding of expanded sizes of both the rapidly exchanging (pool
A) and the slowly exchanging (pool B) pools of body cholesterol
(Table VIII). Recent work reported by Vahouny et al (20) suggested
that in rats there was reduced clearance of both cholesterol and
triglycerides associated with chylomicron-like lipoprotein particles
in animals fed casein as compared with those fed soy protein.
Addition of lysine to soy protein had only slight effects but
addition of arginine to casein led to results similar to those found
with feeding soy protein. In this study, serum insulin levels were
correlated inversely with serum lipid concentrations. This is at
odds with the report of Sugano et al (15) who reported alterations
of glucagon, but not insulin, with addition of arginine to casein.
Both of these reports agree that feeding soy protein resulted in
less cholesterolemia than casein and that total serum cholesterol
was only slightly affected by addition of lysine or arginine. The
differences between the two investigations may be attributable to
differences in age or strain of rats, or different preparations of
soy and casein.

   Whatever the mechanism of hypocholesterolemia associated with
feeding vegetable proteins when compared with most animal proteins,
studies in humans indicate a similar effect only when the initial
serum cholesterol concentration is relatively high. Individuals
with normal cholesterol levels do not respond with a further
decrease in serum lipids. Sirtori et al (21) treated type II
hyperlipoproteinemic patients with a diet that was similar in
composition to the low fat, low cholesterol diet recommended by the
American Heart Association and compared this with a diet which was
similar in composition except for substitution of the majority of
animal protein with textured vegetable protein in a crossover
design. The soy protein diet resulted in a significant reduction in
levels of plasma total and LDL cholesterol, and this decrease was
far greater than with the usual therapeutic diet. When patients
were given the soy diet and then switched to the
usual diet (based on meat as the protein source), there was an
increase of total and LDL cholesterol. This study had been
criticized because of significant differences in the
polyunsaturated:saturated (P/S) fat ratio between the two diets.
Therefore, this group of researchers performed another study in
which the P/S ratio was 0.1 or 2.7 in a crossover design (22). The
P/S ratio did affect plasma cholesterol levels but the soy diet
resulted in reductions of plasma cholesterol that always exceeded
15% regardless of the P/S ratio.

   Not all studies of soy protein diets in humans have found
significant reduction of plasma cholesterol. In addition to
differences in diet formulations and soy protein preparations, one
reason that has been proposed for these disparities is that there is
a statistically significant negative correlation between the initial
cholesterol level and the decrease of cholesterol achieved with the
soy protein diet ($r = -0.48$, $p < 0.01$) (23). In other words, those
with normal or slightly elevated cholesterol levels would be
expected to show less change in cholesterol as a result of the soy
diet than individuals with markedly elevated lipids. Sirtori and
associates (23) have followed a number of type II
hyperlipoproteinemic patients who have been maintained on the soy

protein diet for over two years. After at least two months of total substitution of the dietary protein, the patients were instructed to consume at least six meals per week based on the textured soy protein. A decrease in total cholesterol averaging 30% was maintained and there was a small but significant increase in HDL cholesterol levels.

Studies on the hypocholesterolemic mechanism(s) associated with substitution of soy protein for animal protein have provided seemingly conflicting results. Fumagalli et al (24) have reported that type II hyperlipoproteinemic patients treated with the soy diet had no change in fecal steroid excretion nor any difference in half-life of plasma cholesterol. However, Nestel et al (25) found that vegetarians excreted less neutral sterols and bile acids and had reduced synthesis of LDL than omnivores. Taken together, these studies indicate that the hypocholesterolemic effect of plant proteins is not due to dietary fiber or some other nonabsorbable component of the diet. Furthermore, the mode of action of soy protein appears to be dissimilar from pharmacologic agents that affect steroid excretion into the intestinal lumen or bind bile acids to prevent reabsorption.

A number of potential mechanisms are under study by several groups. The insulin/glucagon ratios may control lipid biosynthesis but data on this point in relation to feeding soy protein have also been conflicting. Different molecular weight forms of the hormones may be secreted in response to feeding the different proteins. Digestion of plant and animal proteins may yield absorption of different peptides or amino acids that have biologic effects. To this end, Sirtori (23) found that plasma content of free arginine increased 20% when patients were fed a soy-based diet. Partial digestion products of the protein may bind lipids in the intestine but data on this point are conflicting. Rates of digestion of the proteins may be different, eliciting altered hormonal or other biologic responses. Products of protein digestion may influence apolipoprotein secretion in either the intestine or liver. Of relevance to this point is the hypothesis that the lysine/arginine ratio of a protein determines its atherogenicity. Since lysine and arginine are the amino acids responsible for net binding by the apoprotein molecule, if vegetable protein can alter the apoprotein moities quantitatively or qualitatively this is another potential pathway for hypocholesterolemic activity. We have also suggested that the high L/A ratio of animal protein might inhibit hepatic arginase activity, thus reducing formation of arginine-rich apoproteins which are characteristic of experimentally induced hyperlipidemia (14). It is possible that more than one mechanism may be found for the hypocholesterolemic effect of vegetable protein. Substantial work remains to be done in identification of the hypocholesterolemic mechanism of feeding soy protein. Whether there are general or specific differences in effects on lipid metabolism between animal and plant proteins also remains to be determined.

Acknowledgments

    Supported, in part, by grant HL-03229 and Research Career
Award HL-00734 from the National Institutes of Health and by funds
from the Commonwealth of Pennsylvania. The authors thank the
following individuals who have contributed to the work discussed in
this review: Drs. Jon A. Story, Susanne K. Czarnecki, and Ms.
Shirley A. Tepper.

Literature Cited

1. Ignatowski, A.; Virchows Arch. Pathol. Anat. Physiol. Klin.
    Med. 1909, 198, 248-270.
2. Anitschkow, N.; Chalatow, S. Zentralbl. Alleg. Pathol.
    Patholog.Anat. 1913, 24, 1-9.
3. Meeker, D.R.; Kesten, H.D. Arch. Pathol. 1941, 31, 147-162.
4. Kritchevsky, D.; Tepper, S.A. Life Sci. 1965, 4, 1467-1471.
5. Kritchevsky, D.; Davidson, L.M.; Kim, H.K.; Krendel, D.A.;
    Malhotra, S.; Vander Watt, J.J.; Du Plessis, J.P.; Winter,
    P.A.D.; Ipp, T.; Mendelson, D.; Bersohn, I. Exp. Mol. Pathol.
    1977, 26, 28-51.
6. Kritchevsky, D.; Tepper, S.A.; Czarnecki, S.K.; Klurfeld, D.M.;
    Story, J.A. Atheroscler. 1981, 39, 169-175.
7. Kim, D.N.; Lee, K.T.; Reiner, J.M.; Thomas, W.A. Exp. Mol.
    Pathol. 1978, 29, 385-399.
8. Kim, D.N.; Lee, K.T.; Reiner, J.M.; Thomas, W.A. Exp. Mol.
    Pathol. 1980, 33, 25-35.
9. Barth, C.A.; Pfeuffer, M.; Hahn, G. Ann. Nutr. Metab. 1984,
    28, 137-143.
10. Hamilton, R.M.G.; Carroll, K.K. Atheroscler. 1976, 24, 47-62.
11. Huff, M.W.; Hamilton, R.M.G.; Carroll, K.K. Atheroscler. 1977,
    28, 187-195.
12. Huff. M.W.; Roberts, D.C.K.; Carroll, K.K. Atheroscler. 1982,
    41, 327-336.
13. Terpstra, A.H.M.; Hermus, R.J.J.; West, C.E. In "Animal and
    Vegetable Proteins in Lipid Metabolism and Atherosclerosis";
    Gibney, M.J.; Kritchevsky, D., Eds.; Alan R. Liss, New York,
    1983; pp. 19-49.
14. Kritchevsky, D.; Tepper, S.A.; Czarnecki, S.K.; Klurfeld, D.M.
    Atheroscler. 1982, 41, 429-431.
15. Sugano, M.; Ishiwaki, N.; Nagata, Y.; Imaizumi, K. Br. J. Nutr.
    1982, 48, 211-221.
16. Eisenstein, A.B.; Strack, I; Gallo-Torres, H.; Georgiadis, A.;
    Miller, O.N. Am. J. Physiol. 1979, 236, E20-E27.
17. Vahouny, G.V.; Chalcarz, W.; Satchithanandam, S.; Adamson, I.;
    Klurfeld, D.M.; Kritchevsky, D. Am. J. Clin. Nutr. 1984, 40,
    1156-1164.
18. Kritchevsky, D.; Tepper, S.A.; Kim, H.K.; Moses, D.E.; Story,
    J.A. Exp. Mol. Pathol. 1975, 22, 11-19.
19. Huff, M.W.; Carroll, K.K. J. Lipid Res. 1980, 21, 546-558.
20. Vahouny, G.V.; Adamson, I.; Chalcarz, W.; Satchithanandam, S.;
    Muesing, R.; Klurfeld, D.M.; Tepper, S.A.; Sanghvi, A.;
    Kritchevsky, D. Atheroscler. 1985, 56: 127-137.

21. Sirtori, C.R.; Agradi, E.; Conti, F.; Gatti, E.; Mantero, O.
     Lancet 1977, 1, 275-277.
22. Sirtori, C.R.; Gatti, E.; Mantero, O.; Conti, F.; Agradi, E.;
     Tremoli, E.; Sirtori, M.; Fraterrigo, L.; Tavazzi, L;
     Kritchevsky, D. Am. J. Clin. Nutr. 1979, 32, 1645-1658.
23. Sirtori, C.R.; Noseda, G.; Descovich, G.C. In "Animal and
     Vegetable Proteins in Lipid Metabolism and Atherosclerosis";
     Gibney, M.J.; Kritchevsky, D., Eds.; Alan R. Liss, New York,
     1983; pp. 135-148.
24. Fumagalli, R.; Soleri, L.; Farina, R.; Musanti, R.; Mantero,
     O.; Noseda, G.; Gatti, E.; Sirtori, C.R. Atheroscler. 1982,
     43, 341-353.
25. Nestel, P.J.; Billington, I.; Smith, B. Metabolism 1981, 30,
     941-945.

RECEIVED December 13, 1985

14

Mortality among Seventh-Day Adventists in Relation to Dietary Habits and Lifestyle

Roland L. Phillips and David A. Snowdon[1]

Loma Linda University, School of Medicine, Center for Health Promotion, Loma Linda, CA 92350

This report summarizes 21 years of mortality follow-up for 25,000 California members of the Seventh-day Adventist church. Compared to the general population, Adventists have an exceptionally low risk of fatal lung cancer which is clearly accounted for by their lack of cigarette smoking. They also have a marked reduction in risk of fatal large bowel cancer, coronary disease, stroke, diabetes and nontraffic accidents. Compared to Adventists who heavily use meat the vegetarian Adventists have a substantially lower risk of fatal coronary disease, fatal diabetes and death from any cause, especially among men. Among Adventist men who use few animal products (meat, milk, cheese, eggs) the risk of fatal prostate cancer is one third that of Adventist men who heavily use such products. Moderate use of coffee is associated with an increased risk of fatal large bowel cancer among both sexes and an increased risk of coronary death and all-cause deaths among males. Since the amount of meat and coffee used by Adventists tends to reflect their overall adherence to the prudent practices advocated by the Adventist church, these findings suggest that the Adventist lifestyle may delay premature death from several major causes of death.

Seventh-day Adventists have been officially organized as a church for about 120 years. Within a few years after their organization, Adventist leaders began to advocate particular habits and practices that were felt to promote better health. Smoking and drinking of alcoholic beverages soon became proscriptions which are carefully adhered to by the vast majority of Adventists (99 per cent nonsmokers, 90 per cent nondrinkers). The Adventist church also recommends (but does not require) its members to avoid certain dietary items such as meat, poultry, fish, coffee, tea, other

[1]Current address: University of Minnesota, Division of Epidemiology, Minneapolis, MN 55455.

caffeine-containing beverages, rich and highly refined foods, and hot condiments and spices. These recommendations account for the fact that currently about one-fourth of Adventists are lifetime lacto-ovo-vegetarians, one-fourth have adopted the vegetarian diet sometime in life (usually shortly after joining the church), and one-half of the Adventists currently follow a nonvegetarian diet. This variation results in a population with much wider variation in exposure to meat than one could find in a typical sample of the general population. Adventists who chose to use little or no meat also tend to follow many other nutritional practices that are believed to promote health. Thus, wide variation in many dietary exposures, in a population which is relatively homogeneous in many other characteristics relevant to cancer risk, provides an ideal setting to test dietary hypotheses.

Serious interest in the health status of Adventists initially arose in the late 1950's as part of a surge of research effort to investigate the health effects of tobacco use. Adventists provide an opportunity to evaluate the risk of smoking-related diseases among a group who reportedly abstain from both tobacco and alcohol. Adventists are of particular interest as research subjects because former smokers or heavy meat users usually dropped their habit because of religious reasons, rather than illness, and a sizable proportion of Adventists never initiated smoking or meat use.

Brief Overview of Study Methods

In 1960 investigators at Loma Linda University collected baseline self-administered questionnaire data on demographic, sociologic, medical, dietary and lifestyle characteristics from 25,000 California Adventists age 35 and over. These subjects were carefully followed through 1965 and death certificates were obtained for all deaths so that causes of death could be tabulated. We have recently completed the task of extending the follow-up of these 25,000 subjects through 1980 via a computer-assisted record linkage procedure with the California death certificate file. During this 21-year follow-up period 7,250 subjects have died. Utilizing this updated mortality data, this report will recapitulate and summarize the key findings which have been reported during the 25 years since the study began (1-15).

Mortality among Adventists Versus the General Population

Early in the follow-up period it became clear that Adventists have a very low risk of lung cancer (1) and other fatal diseases strongly related to use of cigarettes or alcohol (mouth cancer, esophagus cancer, respiratory disease, liver cirrhosis, etc.). Table I shows that the risk of fatal lung cancer among Adventists is 25% of the risk among general population subjects of comparable age. The standardized mortality ratio used on this table is a convenient statistic to use in comparing mortality rates of any particular group (Adventists in this case) to the general population while taking into account the fact that the distribution of age (the principle determinant of death) may not be comparable between the two groups. A standardized mortality ratio of 100 for any given cause of death would indicate that the risk of Adventists dying from that cause is

Table I
Age-Standardized Mortality Ratios for Selected Causes
of Death Among 25,000 White California Adventists, 1960-1980

CAUSE of DEATH	STANDARDIZED MORTALITY RATIO (100 x obs/exp)			NUMBER of DEATHS (both sexes)	
	Male	Female	Both	Observed	Expected[+]
LUNG CANCER	16*	47*	25*	75	306.5
LARGE BOWEL CANCER	55*	46*	49*	175	358.0
BREAST CANCER	--	83*	--	182	219.3
PROSTATE CANCER	74*	--	--	94	126.7
ALL OTHER CANCER	58*	59*	59*	723	1227.6
CORONARY DISEASE	42*	44*	43*	2460	5744.6
STROKE	47*	48*	48*	1035	2154.2
DIABETES	45*	39*	40*	116	288.7
TRAFFIC ACCIDENTS	107	113	110	119	108.4
OTHER ACCIDENTS	47*	35*	40*	123	306.9
SUICIDE	46*	80	57*	44	77.0
ALL CAUSES	46*	51*	49*	7250	14756.4

*The probability that the mortality differential between
Adventists and all U.S. whites is accounted for by chance
alone is under 0.05.

+The number of deaths expected if white California Adventists
(age 35 or over) had the same concurrent risk of dying
(within each of 26 sex-age groups) as all whites in the United
States.

equal to persons of the same age and sex in the general population.
Twelve years ago, we were intrigued by the data on certain
causes of death other than lung cancer (Table I). For several of
these causes (large bowel cancer, breast cancer, prostate cancer,
diabetes) the risk among Adventists is substantially below the
general population even though they are unrelated to cigarette and
alcohol use. This suggests that aspects of the Adventist lifestyle
other than abstinence from tobacco and alcohol may account for their
low risk. We are now in the twelth year of a totally new study which
is primarily designed to identify the specific components of the
Adventist's lifestyle which may reduce their risk of acquiring these
diseases. This ongoing study differs from the previous mortality
study in three major ways: (1) the study population is considerably
larger, (2) the baseline data on lifestyle is more detailed, and (3)
the outcome of interest is newly acquired disease events rather than
only fatal disease events.
The other specific causes of death which are shown on Table I
(coronary disease, stroke, accidents and suicide) are known to be
related to alcohol or tobacco use. However, for fatal coronary
disease, which is the major killer in this country, the risk among

Adventists is substantially below the risk in comparable subjects in
the general population who have never smoked (8). In view of the
abstinence from alcohol in Adventists, it is especially surprising to
note that their risk of fatal traffic accidents is about the same as
the risk in the general population, although other accidents appear
to be substantially lower among Adventists. This may partially
reflect the fact that the cause of a fatal traffic accident is often
outside of the victim's control, while other accidents usually reflect the victim's own behavior. It is conceivable that the relatively high risk of traffic accidents among Adventists may partially
be accounted for by their low use of caffeinated beverages.
Furthermore, the one study that has examined the driving habits of
Adventists found that they tend to accrue somewhat more traffic
citations than comparable peers in the general population (16). The
relatively low risk of suicide among Adventist men raises the possibility that the Adventist way of life may provide effective ways of
coping with the extreme stresses of life. However, the fact that
Adventist women are not clearly at lower risk of suicide does not
support this concept, particularly in view of the tendency of women
to take their religion more seriously than men.
     The bottom line of Table I shows that the age-standardized risk
of dying from any cause among Adventists is one-half of the risk in
the general population. This does not mean that Adventists are
immortal in some mystical way. With the exception of a few causes of
death which are caused almost entirely by tobacco or alcohol (lung
cancer, liver cirrhosis, etc.), the proportion of deaths attributed
to each of the major causes of death in the U.S. is essentially equal
among Adventists and U.S. whites. Adventists simply die later. This
relative delay in the age at which Adventists succumb to most fatal
diseases results in a lower age-standardized death rate and a longer
life expectancy. Compared to their general population peers, life
expectancy among Adventists who reach age 35 is 6 years longer for
men and 4 years longer for women (3). Our ongoing study will enable
us to determine whether this increased life expectancy among
Adventists is due to a delay in acquiring life threatening diseases
or to a tendency to live longer with their diseases. The latter
possibility is not a desirable situation. Longer life is only
desirable if the quality of life is also extended.

Lifestyle Versus Selection

Although comparisons between Adventists and the general population
are interesting, considerable caution is warranted in drawing
inferences from such comparisons. Proponents of the Adventist
lifestyle tend to use such data to extol the virtues of the Adventist
lifestyle which "obviously" accounts for their favorable mortality
rates. It is important to note that over 50% of the Adventist
subjects in this study are adult converts to the church and such
converts are clearly not a random sample of the general population.
In fact, it is reasonable to assume that adults in the general
population who choose to join the Adventist church are a select group
of rather unique and unusual people whose risk of acquiring or dying
from any given disease may have been distinctly different from the
general population at the time they converted to Adventism. If this
were true, the primary explanation for their low risk may not be the

lifestyle they adopted after joining the church, but rather the
nonlifestyle traits which characterize persons who choose to convert
to Adventism. However, it is possible that some converts to
Adventism were living a lifestyle somewhat comparable to Adventists
prior to joining the church. Furthermore, mortality data for the
general population is based on all persons in the population. Thus,
general population mortality data includes deaths among subjects who
were institutionalized or ill at the onset of the study, whereas such
subjects in the Adventist population would not be likely to complete
a questionnaire which is the prerequisite for inclusion in the
Adventist mortality data. Even the lifetime Adventists are not free
from selection bias. Through no choice of their own they were
endowed with multiple characteristics from their parents who were
likely converts to the church. Clearly, the general population is
not a very appropriate comparison group for Adventists. Thus, only
very limited inferences can be derived from such comparisons.

Mortality and Lifestyle Variations Within Adventists

Because of these unavoidable problems in comparing Adventists with
the general population we have adopted the point of view that much
more can be learned about the relationship between lifestyle and risk
of disease by comparing Adventists with Adventists. The varying
degrees of compliance which Adventists exhibit toward the health
recommendations of the church produces a sizable variation in
lifestyle patterns. In fact, for many habits the variation among
Adventists is much greater than one could find in the general
population. For example, approximately half of the Adventists in
this study are lacto-ovo-vegetarians and 20% use meat four or more
days per week. Thus, Adventists provide a rich resource for
investigating the health effects of lifestyle variations without the
need for an external comparison group which would likely confuse the
picture. The remainder of this report will focus primarily on
comparisons between Adventists who have varying degrees of adherence
to the church's recommendation against the use of meat and coffee.

Meat Use. On Table II, meat use refers to the combined use of both
meat and poultry because they were unfortunately combined in a single
question on the questionnaire. Table II utilizes relative risk as an
easily interpetable means of relating meat use to the risk of dying
of several common diseases which are often considered to be lifestyle
related. Subjects who consume no meat are used as the reference category and thus their relative risk is arbitrarily set to one. The relative risk among subjects in other categories of meat consumption re-
lates directly to the reference category. For example a relative risk
of 2.0 for accidental death among subjects who consume meat 4 or more
days per week (hypothetical data not shown on table) would indicate
that the risk of a fatal accident among heavy meat users is twice
that of subjects who use no meat. A relative risk of 0.5 would indicate that the risk in the comparison group is one half that of the
reference group. All relative risks are age-standardized which essentially eliminates the possibility that the risk differential between
the comparison and reference group could be due to a different age
distribution in the two groups. If the measured risk between the
comparison and reference group were solely due to a markedly older

Table II
Relationship of Meat[#] Consumption to the Age-Standardized
Relative Risk of Dying from the Indicated Cause Among
25,000 California Adventists, 1960-1980

| CAUSE of DEATH | SEX | R E L A T I V E   R I S K (in reference to nonusers) | | | NUMBER OF DEATHS | | |
		No Meat (1)	Meat 1-3 days per week (2)	Meat 4+ days per week (3)	(1)	(2)	(3)
LARGE BOWEL CANCER	B	1.0	1.4	0.9	77	44	18
BREAST CANCER	F	1.0	1.2	1.2	92	53	41
PROSTATE CANCER	M	1.0	1.2	1.5	57	25	15
CORONARY DISEASE	M	1.0*	1.4	1.7	607	306	188
	F	1.0*	1.2	1.2	917	395	240
STROKE	M	1.0	1.3	1.2	199	90	37
	F	1.0	1.2	0.9	425	168	83
DIABETES	M	1.0*	1.3	3.6	18	9	13
	F	1.0*	1.2	2.1	39	18	21
ALL CAUSES	M	1.0*	1.2	1.5	1583	719	453
	F	1.0	1.1	1.1	2577	1089	695

*The probability that chance alone accounts for the progressive
increase in relative risk across meat use categories is under
0.05 (under 0.001 for all causes).

[#]Meat refers to combined use of meat and poultry.   There were
no subjects who used fish without also reporting use of meat
and poultry.

age distribution in the comparison group the "raw" (nonstandardized)
relative risk in the comparison group would be substantially greater
than one, while the age-standardized relative risk would be exactly
equal to one.   The "no meat" category includes 1049 subjects who
report occasional meat use (less than once per week).   They were
combined with subjects who indicated their use of meat as "0" days
per week.   All subjects in the occasional use category took the
effort to write a note to this effect on their questionnaire.   We
suspect many more subjects who really belong in the occasional use
category simply recorded their use as "0" because it approximates
their use closer than "1 day/wk".

The most surprising observations on Table II are the lack of a
clear relationship between meat use and risk of fatal cancers of the
large bowel and breast.   Evidence from several other studies suggests
that a direct positive association should be evident (10).   However,
data from a growing number of recent studies are not consistent with
an association between meat use and risk of large bowel cancer or
breast cancer (17-19).   Meat use is also unrelated to all types of
cancer combined.

An association between meat use and fatal prostate cancer is evident in Table II, although it is not statistically significant. Additional analyses (not shown on table) show a similar degree of association between use of other animal products (eggs, milk, cheese) and fatal prostate cancer. If use of all four animal products are combined into an index which reflects their combined use, a very clear relationship emerges (Table III). Thus, it appears that heavy use of all four animal products is associated with a three-fold increase in risk of fatal prostate cancer.

In males the risk of fatal coronary disease shows a substantial stepwise increase with increasing meat use. Although the association between meat use and fatal coronary disease is statistically significant in females, the degree of increased risk with heavy meat use is rather small. The pattern is rather similar for both males and females when all causes of death are considered together. However, these relationships with meat use are somewhat stronger among younger males under age 65 (Table IV). In this age group of males, the risk of fatal coronary disease among heavy meat users is increased by a factor of 2.5. Furthermore, heavy meat use is associated with a 40% increase in the risk of dying of noncoronary causes and a 70% increase for all causes. Thus, causes of death other than coronary disease substantially contribute to this increased risk of dying.

The rather strong relationship between meat use and fatal diabetes was only recently discovered, and to our knowledge this relationship has not been reported by other investigators. Since most people with diabetes do not die from their disease, mortality data is not the preferable type of data to use in investigating a possible causative link between diabetes and meat use. However, the relationship is fairly strong and shows a dose-response pattern for both males and females despite a very limited number of diabetic deaths. There is a clear need for further studies to evaluate the relationship of meat and other dietary habits to the frequency of diabetes utilizing more direct methods of measuring diabetes risk.

Coffee Use. Table V focuses on the relationship of coffee use to the same six causes of death considered on previous tables. Approximately 70% of the Adventists in this study use no coffee. However, 17% drink two or more cups of coffee per day and 10% limit their intake to one cup per day. The "no coffee" category includes 362 subjects who report occasional coffee use (less than one cup per day). They were combined with subjects who indicated their coffee use as "0" cups per day. All subjects in the occasional use category took the effort to write a note to this effect on their questionnaire. We suspect many more subjects who really belong in the occasional use category simply recorded their use as "0" cups per day because it approximates their use closer than "1 cup per day".

Unfortunately, the questionnaire did not distinguish between caffeinated versus decaffeinated coffee or instant versus brewed coffee. Positive associations with coffee use are seen for fatal large bowel cancer in both sexes, fatal coronary disease in males only, and possibly for all causes of death in males only. Both sexes were combined for large bowel cancer because the degree of association was essentially equivalent when examined separately for

Table III
Relationship of Animal Product Use to the Age-Standardized
Relative risk of fatal Prostate Cancer Among 9000
California Adventist Males, 1960-1980

ANIMAL PRODUCT INDEX[+]	RELATIVE RISK OF FATAL PROSTATE CA	NUMBER of DEATHS
<5	1.0	15
6	1.2	12
7	1.2	16
8	1.5	20
9	1.8	19
10+	2.9	18

*The probability that chance alone accounts for the progressive increase in relative risk as the animal product index increases is 0.002.

[+]A score of 1,2, or 3 was respectively assigned to 3 categories which reflect increasing use of each of four animal products (meat, eggs, milk, and cheese). For each subject, the scores for these 4 animal products were summed to obtain the animal product index.

Table IV
Relationship of Meat Consumption to the Age-Standardized
Relative Risk of Death From Coronary Disease and All
Other Causes Among California Adventist Men age 35-64,
1960-1980

CAUSE of DEATH	RELATIVE RISK (in reference to nonusers)			NUMBER of DEATHS		
	No Meat	Meat 1-3 days per week	Meat 4+ days per week			
	(1)	(2)	(3)	(1)	(2)	(3)
CORONARY DISEASE	1.0*	1.7	2.5	78	73	73
OTHER CAUSES	1.0*	1.4	1.4	176	137	94
ALL CAUSES	1.0*	1.5	1.7	254	210	167

*The probability that chance alone accounts for the progressive increase in relative risk across categories of meat use is under 0.05.

Table V
Relationship of Coffee Consumption to the Age-Standardized
Relative Risk of Dying from the Indicated Cause Among
25,000 California Adventists, 1960-1980.

CAUSE of DEATH	SEX	RELATIVE RISK (in reference to nonusers)			NUMBER of DEATHS		
		No Coffee (1)	Coffee one cup per day (2)	Coffee 2+ cups per day (3)	(1)	(2)	(3)
LARGE BOWEL CANCER	B	1.0*	1.2	1.7	91	14	31
BREAST CANCER	F	1.0	1.1	0.9	131	19	26
PROSTATE CANCER	M	1.0	0.8	0.7	76	7	11
CORONARY DISEASE	M	1.0*	1.6	1.5	716	127	220
	F	1.0*	1.1	1.2	1111	49	216
STROKE	M	1.0	1.2	0.9	242	32	41
	F	1.0	1.4	0.9	477	84	72
DIABETES	M	1.0	0.7	1.8	26	2	10
	F	1.0	1.4	0.9	56	10	9
ALL CAUSES	M	1.0*	1.4	1.3	1873	282	510
	F	1.0	1.2	1.1	3069	463	591

*The probability that chance alone accounts for the progressive increase in relative risk across coffee use categories is under 0.05 (under 0.001 for all causes)

males and females. However, due to the limited number of large bowel cancer deaths, statistical significance was achieved only when both sexes were combined. The association of coffee use with fatal large bowel cancer was rather unexpected. Although one other study has reported a similar positive association (17), there are two studies that report a negative association suggesting that risk of large bowel cancer is lowered by heavy coffee use (20,21). It seems prudent to interpret our data cautiously at this point. We will re-evaluate this relationship in our new ongoing study and hopefully it will also be studied by other investigators. The lack of a clear dose-response effect in the observed association between coffee use and fatal coronary disease in men somewhat weakens the possibility that the association is causative, however, it is statistically significant and cannot be ignored. The same can be said for the possible relationship between coffee use and the risk of dying from any cause.

One of the difficulties encountered in evaluating the possible associations between disease risk and consumption of meat or coffee is the fact that Adventists who use one of these products also tend to use the other. For example, 91% of the Adventists in this study who use no meat also abstain from coffee, whereas, 50% of those who use meat four or more days per week also use coffee. Thus, it is

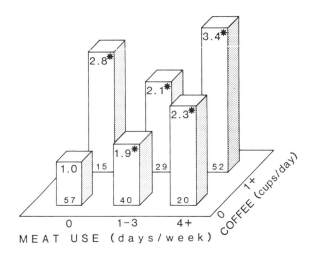

Figure 1. Age-standardized relative risk of fatal coronary
disease among California Adventist males age 35-69 divided by
meat and coffee use, 1960-80. The number of deaths is indicated
at the base of each bar. *The probability that chance alone
accounts for the differential in risk between the indicated group
and subjects who use no meat or coffee is less than 0.001.

quite possible that associations between meat use and disease could
be reflecting the effect of coffee, and vice versa. Unfortunately,
there are too few deaths for most specific causes to reliably
disentangle the separate effect of meat and coffee. However, for
fatal coronary disease there are adequate deaths to partially assess
the possibility of independent associations with meat and coffee.
Figure 1 depicts these associations for fatal coronary disease in
young males. The figure is limited to relatively young males because
this group showed the strongest associations when meat and coffee
were examined separately. The important trends to note are: (1) the
stepwise increase in risk with increasing meat use among subjects who
use no coffee (front row of bars), (2) the near threefold risk
increase of coffee users over nonusers among subjects who use no meat
(left row of bars), and (3) the coffee users who also use meat
heavily have the highest risk (3.4 times the risk among subjects who
use neither meat or coffee). The pattern rather strongly suggests
that both meat and coffee use are substantially associated with an
increased risk of fatal coronary disease among young males. The
pattern for all causes of death among young males was rather similar
to coronary disease (no data shown) with the exception that coffee
use was more strongly associated with risk of dying than meat use.

Summary and Comments

The following summary statements deserve consideration and comment:
1. Compared to the general population, Adventists have an exceptionally low risk of fatal lung cancer and other diseases which are strongly related to cigarette or alcohol use. They also appear to have a marked reduction in risk of death from large bowel cancer, coronary disease, stroke, diabetes, and nontraffic accidents.
2. Meat use among Adventists is unrelated to fatal cancers of the large bowel and breast.
3. Animal product use among Adventists is strongly related to risk of fatal prostate cancer.
4. In Adventist males, meat use is strongly related to risk of death from coronary disease and all causes, especially in younger males.
5. Meat use among Adventists is strongly related to fatal diabetes among both males and females.
6. Meat use among Adventists is questionably related to coronary disease in females, and stroke in males.
7. Coffee use among Adventists is moderately related to risk of death from large bowel cancer in both sexes, as well as coronary disease and all causes in males.
8. Coffee use and meat use are both independently related to risk of fatal coronary disease in young Adventist males.

Overall, it does appear that the vegetarian diet, abstinence from smoking, and perhaps other prudent practices adopted by the majority of Adventists may be preventive for several of the major life threatening diseases which are so prevalent in the U.S.

The most important item to keep in mind when interpreting this data is that all the relationships mentioned are merely associations between a disease outcome and some personal characteristic which is common to a high proportion of subjects who experience the disease. Even if statistical testing has essentially ruled out chance phenomenon as a likely explanation for these observed associations, there is still the very real possibility that the associations are indirect and, thus, not directly relevant to the cause of the disease. For example, it is likely that Adventists who use meat and/or coffee may have many other characteristics which are different from subjects who abstain from these products. One or more of these characteristics may be the important factor which actually accounts for the association between meat and a specific cause of death. Yet, such a factor may not have been measured or taken into account during the data analysis.

It is beyond the scope of this report to pursue a detailed discussion of such factors. However, for all the strong associations mentioned in this report we have identified other items on the questionnaire which could possibly account for the observed associations. We have re-evaluated each association after taking such items into account. Without presenting the evidence, I will simply state that all associations in this report remained essentially unchanged after the above procedures. Of course it is impossible to take items into account which were not measured by the baseline questionnaire. Preliminary analyses of the baseline data from our new ongoing study strongly suggests that amount of meat use, and to some extent coffee use, is a rather accurate index of the degree of adherence to multiple aspects of the Adventist lifestyle including degree of involvement with church activities. Thus, it is possible that many of the associations in this report primarily reflect the effects of adherence or nonadherence to the total life-style patterns the Adventist church has advocated for over 100 years.

One important criterion for determining whether an association is causative is whether it is biologically plausible. Although a justification of the biological plausibility of each of our observed associations is beyond the scope of this report, it is rather common knowledge that abundant evidence from other studies confirms the plausibility of a causative link between meat use and coronary disease (22). However, this study is the first major observational study to clearly show this relationship among U.S. subjects. The associations between meat and diabetes are certainly plausible, but there is considerably less evidence from other experimental or observational studies to substantiate such a relationship.

In reference to coffee use and large bowel cancer, biological plausibility remains to be evaluated. However, it is worth noting that all types of coffee contain substances which are capable of damaging genes (23), and some of the nonliquid portion of ingested coffee does come into direct contact with the inner lining of the bowel wall where cancer begins.

We suspect that the magnitude of most of the associations noted between meat or coffee and specific fatal diseases are somewhat underestimated because Adventists may tend to underreport the amount of meat or coffee they use. If a substantial number of subjects actually use more meat and coffee than they reported on the initial questionnaire, it would tend to make it harder to find the real associations, and the observed associations would tend to be weaker. Furthermore, we may have missed associations because subjects changed their habits during the 21-year follow-up period. All observed associations are based on meat and coffee use at the time subjects completed the baseline questionnaire (1960). Subsequent changes in these habits would tend to reduce or eliminate the possibility of finding disease associations with these habits. Failure to find associations, or detection of weak associations, could also result from the fact that our study population contains relatively few subjects who are very heavy users of meat or coffee, while it contains an abundance of subjects who have no exposure to these items.

There are many aspects of the Adventist lifestyle that have not been considered in this report. However, if we assume that the positive associations noted in this report are conservative, it is reasonable to conclude that some aspects of the Adventist lifestyle may substantially reduce the risk of premature death.

Literature Cited

1.   Lemon, F. R.; Walden, R. T.; Woods, R. W. Cancer 1964, 17, 486-97.
2.   Lemon, F. R.; Walden R. T. J. Am. Med. Assoc. 1966, 198, 117-26.
3.   Lemon, F. R.; Kuzma J. W. Arch. Environ. Health 1969, 18, 950-55.
4.   Phillips, R. L. Cancer Res. 1975, 35(suppl), 3513-22.

5.   Phillips, R. L.; Lemon, F. R.; Beeson, W. L.; Kuzma, J. W. Am.
     J. Clin. Nutr. 1978, 31(Suppl), 191-98.
6.   Phillips, R. L. J. Environ. Path. Toxicol. 1980, 3, 157-69.
7.   Phillips, R. L.; Garfinkel, L; Kuzma, J. W.; Beeson, W. L.;
     Lotz, T. L.; Brin, B. J. Nat. Cancer Inst. 1980, 65, 1097-1107.
8.   Phillips, R. L.; Kuzma, J. W.; Beeson, W. L.; Lotz, T. Am. J.
     Epidemiol. 1980, 112, 296-314.
9.   Phillips, R. L.; Snowdon, D. A. Cancer Res. 1983, 43(Suppl),
     2403-8.
10. Phillips, R. L.; Snowdon, D. A.; Brin, B. N. In "Environmental
     Aspects of Cancer - The Role of Macro and Micro Components of
     Foods"; Wynder, E. L.; Leveille, G. A.; Weisburger, J. H.;
     Livingston, G. E., Eds.; Foods and Nutrition Press: Westport,
     Conn., 1983; pp. 53-72.
11. Snowdon, D. A.; Phillips, R. L. Am. J. Public Health 1984, 74,
     820-23.
12. Snowdon, D. A.; Phillips, R. L.; Choi, W. Am. J. Epidemiol.
     1984, 120, 244-50.
13. Snowdon, D. A.; Phillips, R. L.; Fraser, G. E. Prev. Med.
     1984, 13, 490-500.
14. Kahn, H. A.; Phillips, R. L.; Snowdon, D. A.; Choi, W. Am. J.
     Epidemiol. 1984, 119, 775-87.
15. Snowdon, D. A.; Phillips R. L. Am. J. Public Health 1985, 75,
     507-12.
16. Kuzma, J. W.; Dysinger, P. W.; Strutz, P.; Abbey, D. Accid.
     Annaly. Prev. 1973, 5, 55-65.
17. Graham, S.; Dayal, H.; Swanson, M.; Mittelman, A.; Wilkinson, G.
     J. Nat. Cancer Inst. 1978, 61, 709-14.
18. Hirayama, T. In "Gastrointestinal Cancer-Endogenous Factors";
     Bruce, W. R.; Correa, P.; Lipkin, M.; Tannenbaum, S. E.;
     Wilkins, T. D., Eds.: Cold Spring Harbor Laboratory: New York,
     1981; pp 409-26.
19. Graham, S.; Marshall, J.; Mettlin, C.; Rzepka, T.; Nemoto, T.;
     Byers, T. Am. J. Epidemiol. 1982, 116, 68-75.
20. Haenszel, W.; Berg, J. W.; Segi, M.; Kurihara, M.; Locke, F. B.
     J. Nat. Cancer Inst. 1973, 51, 1765-79.
21. Bjelke, E. Ph.D. Thesis, Univ. of Minnesota, Minneapolis, 1973.
22. Blackburn, H. Prev. Med. 1983, 12, 2-10.
23. Nagao, M.; Takahashi, Y.; Wakabayashi, K.; Sugimura, T. Mutation
     Research 1979, 68, 101-6.

RECEIVED December 26, 1985

CHEMISTRY–COMPOSITION

15

Composition and Functionality of Protein, Starch, and Fiber from Wet and Dry Processing of Grain Legumes

F. W. Sosulski[1] and K. Sosulski[2]

[1] Department of Crop Science and Plant Ecology, University of Saskatchewan, Saskatoon, Saskatchewan, Canada S7N 0W0
[2] Biomass Resources, Saskatchewan Research Council, Saskatoon, Saskatchewan, Canada S7N 0X1

The field pea (Pisum sativum) and small fababean
(Vicia faba minor) were pin milled and air
classified into protein and starch fractions or,
alternately, the protein, starch and fiber were
extracted by an aqueous alkali procedure.
The efficiencies of protein (75-80%)
and starch (88-93%) recoveries by the dry
process were higher than the 73-79% recoveries
by wet processing, and there were no losses
of solids in the whey and wash water or
need for effluent recovery. The starch fraction
was similar to refined starch in most functional
properties except for a low amylograph viscosity.
The protein fraction showed low nitrogen solubility
and rather low water hydration and oil absorption
values relative to those of the proteinates but
oil emulsification was quite high. Refined legume
fiber had a water hydration capacity of over 20 g/g
product.

Characteristically, legume seeds are rich in protein and contain
intermediate to high levels of lysine and threonine which are
important in balancing the deficiencies of these essential amino
acids in cereal diets. Certain legume proteins, such as soybean,
also exhibit strong functional properties, especially water
solubility, water and fat binding and emulsification. Thus soybean
flours, protein concentrates and isolates have been used widely as
nutritional supplements and functional ingredients in foods.
    Most grain legumes are consumed primarily as whole or split
seeds, and only limited quantities are processed into flours or more
refined products. Starchy legume flours appeared to have weaker
functional properties than defatted soybean or lupine flours (1),
due in part to their lower protein contents. To overcome the
problem of variable protein composition among samples of field peas
and other grain legumes, Youngs (2) developed a process for
separation of the protein and starch fractions in the seeds by fine
grinding and air classification. The fine fraction, obtained in

0097-6156/86/0312-0176$06.00/0
© 1986 American Chemical Society

yields of about 25%, contained 48-67% of protein, depending upon the legume species (3). The functional properties of oil absorption, emulsification and whippability in the air-classified protein fractions were greatly enhanced over the flours. The oligosaccharides, trypsin inhibitors, hemagglutinins, saponins, phytic acid and, in fababean, vicine and convicine concentrated into the fine fraction with the protein (4,5). Numerous other investigators have evaluated the fractions obtained by air classification of finely-ground flours from field pea, fababean and other starchy legumes (6,7,8,9).

Grain legumes have also been processed into refined starch (10,11) and protein isolates (12,13,14) by procedures derived from the traditional corn starch and soybean protein industries (15). However, comparative data on product yields, composition and losses have not been published. A commercial plant for the wet processing of field pea into refined starch, protein isolate and refined fiber has been established in Western Canada. Little is known about the characteristics of the protein isolate or refined fiber product. Water-washed starch prepared from the air-classified starch fractions of field pea (16,17) and fababean (6) have been investigated for certain physico-chemical and pasting properties. Reichert (18) isolated the cell wall material from soaked field pea cotyledons and determined its fiber composition and water absorption capacity. In addition, the effects of drying techniques on the characteristics of pea protein isolates have been determined (14).

The objectives of the present study were to compare the processes of protein and starch concentration by dry air classification and wet alkali extraction of protein and starch from field pea and fababean. The yields, composition and functionality of the crude and refined products were compared.

Experimental Procedures

The smooth field pea (Pisum sativum L. cv. Trapper) and small fababean (Vicia faba minor cv. Diana) samples were composites of several lots grown in 1981. After blending, about 20 kg were dehulled on a resinoid disc, abrasive dehuller (19), followed by air aspiration to remove 10 and 15% of hulls, resp. The dehulled seeds were coarse-ground in a hammermill and subdivided for dry and wet processing.

Dry Process. Ten kg each of the ground field pea and fababean were passed through an Alpine Pin Mill model 250 CW (Alpine American Corp., Natick, MA) (Figure 1). Two passes through the mill reduced the particle size to less than 325-mesh. The pin-milled flours were then fractionated into light and dense particles by a single pass through the Alpine Air Classifier Type 132 MP at a cut point of 15 microns (800-mesh) diameter between the two fractions (9), followed by a reclassification of the dense fraction (20). The two protein fractions were combined.

Wet Process. About 5 kg of each meal were slurried in 25 l of 0.02% NaOH for 1 hr at room temperature. The mixture was stirred with a high speed mixer (Polytron) until a uniform slurry was obtained (Figure 2). The yellowish suspension was passed through a

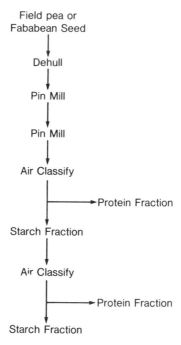

Figure 1.   Flow diagram for pin milling and air classification of field pea and fababean into protein and starch fractions.

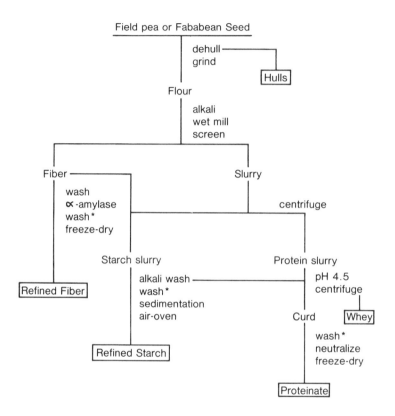

* Discarded

Figure 2.   Flow diagram for preparation of purified fiber, starch and proteinate from field pea and fababean by wet processing techniques.

Morehouse 200 wet mill (Morehouse Industries, Fullerton, CA),
operated at 3600 rpm. The finely dispersed slurry was filtered
through a 230-mesh vibrating screen to remove the coarse fiber
fraction. The residue was washed a second time with 0.02% NaOH for
1 hr and passed through the wet mill and screen to remove additional
solubles.

The fiber fraction was washed several times with water, ratio
1:5, the washings being added to the starch fraction (Figure 2).
Final traces of starch were removed from the fiber by incubation
with 2% alpha-amylase (Sigma Chemical Co., type V-A from hog
pancreas) for 16 hr at 37°C, and washing several times with
distilled water before freeze-drying.

The final pH of the initial extract from the vibrating screen
was 7.5 but this increased to pH 9.6 and 10.2 for fababean and field
pea, resp., during the second extraction. The filtrates were
combined and centrifuged on a Fletcher basket centrifuge at 3400 rpm
(1800 x g) to separate the starch and protein (Figure 2).

The starch fraction was washed initially with 0.02% NaOH, the
extract being added to the protein solution (Figure 2). The starch
was then slurried in distilled water, separated by sedimentation,
and dried at 30°C. The combined protein extracts were adjusted to
pH 4.5 with 1N HCl and the whey separated from the curd by
centrifugation in the basket centrifuge (1100 x g). The protein
curd was washed twice with water adjusted to pH 4.5, then
resuspended at pH 7.0 using 1N NaOH. The proteinate was
freeze-dried.

Analytical Methods. The samples were analyzed by standard AACC (21)
procedures for moisture (air-oven method), protein (Method 46-13),
crude fat (Method 30-25), crude fiber (Method 32-10), insoluble
dietary fiber (Method 32-20) and ash (600°C, 3 hr). Starch
content was determined by the polarimeter method (Method 76-20) and
total sugars by Method 80-60. Color characteristics of the dried
products were evaluated with the Hunter Color Difference Meter.

Functional property tests were conducted in duplicate. AACC
(21) methods were used for the determination of water hydration
capacity (Method 88-04) and nitrogen solubility index (NSI) (Method
46-23). Oil absorption capacity was measured by the procedures of
Lin et al. (22) and oil emulsification by a modification (22) of
the Inklaar and Fortuin (23) method. Pasting characteristics of
12.0% (w/v, db) slurries of the flours and processed products were
determined on a Brabender Visco/Amylograph (Method 22-10). The
slurries were heated from 30 to 95°C before cooling to 50°C to
obtain the cold paste viscosity value. Gelation experiments were
conducted by heating 15% (w/v db) slurries in sealed stainless steel
containers to 90°C for 45 min in a water bath (3).

Results and Discussion

Composition of legume flours. Proximate analyses of the dehulled
flours of field pea and fababean (Table I) showed that the raw
materials were typical of legume flours processed by previous
investigators (3,8,24) but Colonna et al. (6) and Vose et al. (9)
utilized field pea samples with higher concentrations of crude

Table I.   Concentration of proximate constituents and carbohydrates
in flours, dry and wet processed products from dehulled
field pea and fababean, % dry basis.

Process and Product	Crude protein*	Crude fat	Crude fiber	Ash	Starch	Other	Total sugars
Field peas: Flour	22.2	1.3	1.3	2.8	55.0	17.4	7.0
Dry-Protein fraction	52.7	2.9	2.9	5.7	8.3	27.5	12.7
Starch fraction	6.4	0.6	0.5	1.3	83.2	8.0	3.5
Wet-Proteinate	87.7	3.0	0.2	5.8	0	3.3	–
Refined starch	0.4	0.1	0.8	0.2	94.0	4.5	–
Refined fiber	0.8	1.2	47.3**	1.7	0	49.0	–
Fababean: Flour	33.5	1.0	1.1	3.2	51.3	9.9	4.8
Dry-Protein fraction	72.5	2.6	2.0	5.7	8.6	8.6	7.1
Starch fraction	9.9	0.5	0.6	1.6	77.3	10.1	4.2
Wet-Proteinate	94.1	1.9	0.1	3.1	0	0.9	–
Refined starch	0.5	0.1	0.6	0.2	94.1	4.5	–
Refined fiber	0.7	1.2	36.9**	2.2	0	59.0	–

* N x 6.25.
**Insoluble dietary fiber.

protein. In the present study, the concentrations of crude protein
and starch of 22.2% and 55.0%, resp., in field pea contrasted with
the 33.5% crude protein and 51.3% starch in fababean. The concentrations of crude fat, crude fiber and ash totalled only about 5%e of
the legume flours. Th "other" component was calculated by difference and, since the flours differed widely in this value, analyses
for total sugars were done. The 7.0% and 4.8% of total sugars in the
two flours represented 40–50% of the unknown component. The remainder would be phytate (5), oligosaccharides (4) and components of
dietary fiber lost during the crude fiber analysis (18).

Moisture in the original field pea and fababean seeds obtained
from storage varied from 11.2% to 12.7%. During pin milling, moisture losses were quite high, about one-third of the seed moisture,
and air classification reduced the value further to about 6.0% in
both the protein and starch fractions. During storage the refined
starches equilibrated to about 11.8% moisture whereas the freeze-
dried proteinate and fiber were at 6.7% and 4.4% moisture, resp. To
compare the yields and compositions on an equal basis, the results
are reported on a dry weight basis.

Dry and Wet Process Efficiencies. The yields of products obtained by
the two processes are reported in grams per kilogram of raw material
(Table II) and as percentages of the original flour weights (Figure
3). The compositions of the protein, starch and fiber components are
given in Table I and the distribution of protein and starch among the
fractions are given in Figure 3.

The cut point of 15 microns on the Alpine air classifier gave an
fine:coarse split of 31.8:61.7 for field pea but the yield of protein
fraction was much higher at 37.0% for fababean (Table II). Based on
the protein contents of the fine fractions (Table I), the recoveries
of protein in the fine fraction were 75.5% for field pea and 80.0%

Table II. Yields and processing efficiency of dry and wet milled products from field pea and fababean, dry basis.

Process and product	Protein fraction	Protein-ate	Starch fraction	Refined starch	Refined fiber	Whey solids	Wash solids losses	Invisible losses	Recovery of Protein	Starch
			g product/kg raw material						%	%
Dry process										
Field pea	318	–	617	–	–	–	–	65	75.5	93.3
Fababean	370	–	561	–	–	–	–	69	80.0	87.7
Wet process										
Field pea	–	182	–	484	97	186	107	–	72.7	79.2
Fababean	–	265	–	391	68	185	111	–	74.2	76.2

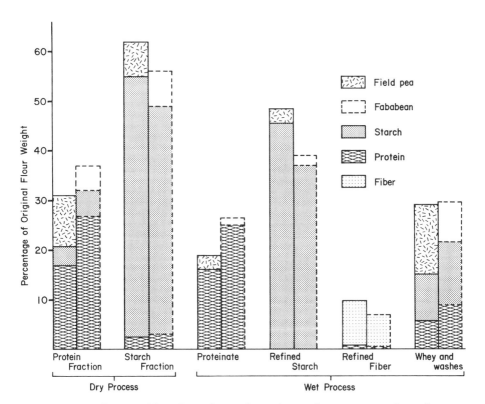

Figure 3. Yields of products from dry and wet processing of field pea and fababean flours and their protein and starch concentrations in percent.

for fababean (Table II). Residual starch in the protein fractions
was only 8.3 to 8.6%, so that recoveries of starch in the coarse
fractions averaged 90% in the two legumes. Figure 3 illustrates
that both high starch (field pea) and high protein (fababean)
legumes can be efficiently fractionated by the air classification
technique. However the field pea protein fraction had a much lower
protein concentration than fababean. and the level of non-starchy
carbohydrates was particularly high (Table I).

The dry process technique used in the present study involved
passing the legume flours twice through the pin mill before the
double pass through the air classifier. The efficiencies of protein
and starch separation into the fine and coarse fractions (Table II)
were greater than was obtained by Sosulski and Youngs (3) when
single passes through the pin mill and air classifier were employed.
Tyler et al. (8,20) reported higher protein and starch separation
efficiencies by the double milling and classification procedure.

On the basis of proteinate yields (Table II) and their protein
contents (Table I), the recoveries of protein during wet processing
were about 73% for both legumes, which was only slightly below the
efficiency of the dry process. However, the losses of starch in the
whey and wash solids were substantial. and starch recoveries averaged 77.5%. The yields of refined fiber were about 8% of the raw
materials. Almost 30% of the dry matter from wet processing would
have to be recovered from whey and wash extracts to make the process
economical.

The protein concentration in the field pea proteinate was only
87.7% due to the presence of significant quantities of ash, lipid
and carbohydrate (Table I). Both refined starches were relatively
pure. the protein levels being only 0.5%. The merits of producing
protein and starch isolates as opposed to concentrates by the dry
process would depend on their relative functional properties and the
requirements of the end-user.

The refined fiber obtained during wet processing differed from
the hull material in being white in appearance, arising primarily
from cell wall material in the cotyledons (18). The yield of
refined fiber and its concentration of insoluble dietary fiber from
field peas greatly exceed that of fababean (Table I). It appeared
this component, along with total reducing and non-reducing sugars.
oligosaccharides and phytin, accounted for the high proportion of
"other carbohydrates" in the protein fraction from field pea.

Functional Properties. The pH's of the flours and products obtained
by air classification varied between 6.5-6.7 (Table III), which was
typical of legume flours (1.3). The proteinates were near pH 7 because of the neutralization procedure after isoelectric precipitation while the refined starch and fiber were still alkaline in pH
despite several washings with distilled water. In a previous study,
adjustment of the pH of lupine flour was shown to have a significant
influence on functional properties (1) but pH was not adjusted in
the present investigation.

The field pea and fababean flours, which were ground to
60-mesh. showed high nitrogen solubility (Table III). However, the
process of pin milling reduced the NSI values to approximately
one-half that of the original flours. The air classified protein

Table III   Functional properties of flours, dry and wet processed products from dehulled field pea and fababean, dry basis.

Process and product	pH	Nitrogen solubility index %	Water hydration capacity g/g sample	Oil absorption capacity g/g sample	Oil emulsification %
Field pea: Flour	6.70	82.4	0.88	73.0	64.0
Dry-Pin-milled flour	6.64	45.8	0.80	63.0	79.0
Protein fraction	6.51	47.2	1.22	93.0	76.0
Starch fraction	6.82	-	1.01	58.7	14.0
Wet-Proteinate	6.80	83.6	2.63	249.0	78.0
Refined starch	8.41	-	1.16	68.0	6.6
Refined fiber	8.42	-	20.10	-	-
Fababean: Flour	6.52	84.3	0.66	79.0	55.0
Dry-Pin-milled flour	6.60	39.0	0.58	48.0	74.0
Protein fraction	6.50	43.0	1.04	89.7	70.0
Starch fraction	6.72	-	0.96	55.7	18.0
Wet-Proteinate	7.00	99.0	2.42	171.0	73.0
Refined starch	8.43	-	1.03	64.0	8.0
Refined fiber	8.42	-	22.30	-	-

fractions exhibited similar low NSI values. Exit temperatures of
55-60°C were recorded in the product after pin milling and excessive heating during grinding was likely the main factor responsible
for the low NSI's. The proteinate of fababean was completely soluble under the conditions of the NSI test. The somewhat lower NSI
value for field pea proteinate, 83.6%, indicated that phytin was
present and bound to the protein (5), resulting in a high ash value
for the product as well (Table I).

The water hydration capacities of 0.6-0.9 g water/g of legume
flour were improved to 1.0-1.2 g/g in the protein and starch fractions (Table III). However, the proteinates of field pea and fababean were markedly higher in water hydration capacity, being more
than two times greater than was shown by the corresponding dry-
processed protein fraction. The starch fractions and refined
starches had similar water hydration capacities of 1.0 g/g product.
Reichert (18) reported that cell wall material from field pea cotyledons absorbed approximately 18 times its weight of water. In the
present study field pea and fababean fiber absorbed about 20 and 22
times, resp., of their weight of water. These high values were
attributed to the high proportion of pectic substances in the cell
wall material (18).

Like NSI, the oil absorption capacities of the legume flours
were decreased by pin milling, and the protein fractions were more
functional in this parameter than were the starch fractions

Table IV. Visco/Amylograph and gelation properties of
flours, dry and wet processed products from dehulled
field pea and fababean.

Process and product	Peak viscosity B.U.*	Cold paste viscosity B.U.	Degree of gelation %	Gel colour
Field pea:Flour	400	840	60	cream
Dry-Pin-milled flour	400	785	81	cream
Protein fraction	65	110	100	yellow
Starch fraction	260	600	96	cream
Wet-Proteinate	-**	-	100	brown
Refined starch	610	1250	100	white
Fababean:Flour	340	720	62	cream
Dry-Pin-milled flour	315	715	80	cream
Protein fraction	60	90	94	yellow
Starch fraction	150	500	88	cream
Wet-Proteinate	-	-	100	green
Refined starch	480	1480	100	white

*Brabender units.
**Values less than 15 BU.

(Table III). Again, the proteinates absorbed 2-2.5 times more oil
than the corresponding protein fraction, while the wet- and dry-
processed starch were similar in their low absorption capacity.

Pin milling alone improved one functional property, oil emulsification (Table III). The pin-milled flours, protein fractions and
proteinates gave oil emulsification values of 70.0-79.0% compared
to values of less than 18.0% for starch products.

Viscosity and Gelation Characteristics. The field pea and fababean
flours gave typical legume amylograms (1) with intermediate peak viscosities and positive setback or cold viscosity values (Table IV).
The protein fractions gave low viscosities during the heating and
cooling cycle and proteinates showed almost no change in viscosity.
Although the viscosity properties appeared to be due almost entirely to the presence of starch, the starch fractions failed to develop viscosities comparable to the flours. The refined starches gave
the highest peak and cold paste viscosities but the differences
between field pea and fababean were not consistent.

The proportion of the legume flours which gelled when heated
in a closed container increased from about 60 to 80% as a result of
pin milling (Table IV). Also, most of the protein and starch
fractions from field pea gelled under these conditions but a portion of fababean fractions remained as a pourable slurry. The
proteinate and refined starch gave very firm gels.

Under these conditions of heating the aqueous slurries, the
flours and starch fractions generally developed creamy colours
(Table IV). However the protein fractions were distinctly yellow
after heating and the proteinates became light brown, or green in
the case of fababean. The refined starches showed only a white
colour during the heating cycles employed in these tests.

Product Colours. Both legume flours showed creamy-yellow colours
under the Hunter Color Difference Meter but fababean flour was also
slightly greenish (Table V). Pin milling improved the lightness of
the flour and this colour was retained by the protein and starch
fractions at the expense of the yellow values. The proteinates
were light brown in appearance whereas the refined starches were
essentially white. The refined fiber retained only a light shade
of yellow as compared to the other products.

Conclusions

The wet processing procedures provided protein, starch and fiber isolates which exhibited strong functional properties. However, processing costs would be high due to extensive losses of solids in
the whey and wash waters, and the need to dry the products as well
as recycle the effluents. The pin milling and air classification
system provided less refined products of intermediate protein or
starch composition but losses were negligible and there was no need
for addition or removal of water in the process. Functionally, the
protein fraction showed low NSI values and only marginal improvement in water hydration or oil absorption capacity. Oil
emulsification and gelation values were comparable to those of the
protein isolate. The starch fraction was comparable to refined
starch in most function properties except for amylograph
viscosities.

Table V. Hunterlab colour values (L=lightness; a= +red, -green;
          b= +yellow, -blue) of flours, dry and wet processed
          products from dehulled field pea and fababean.

Process and	Field pea			Fababean		
product	L	a	b	L	a	b
Legume flour	87.9	-0.3	19.1	88.3	-1.6	18.3
Dry process						
Pin-milled flour	92.8	-1.1	13.2	93.4	-1.9	11.5
Protein fraction	92.7	-1.0	13.3	94.3	-1.7	10.4
Starch fraction	93.8	-1.2	11.6	93.5	-2.0	11.0
Wet process						
Proteinate	81.3	0.3	24.2	79.1	-0.9	17.5
Refined starch	96.8	-0.7	2.4	95.9	-0.6	3.2
Refined fiber	91.4	-1.1	8.4	88.2	-0.6	7.9

Literature Cited

1. Sosulski, F.; Garratt, M.; Slinkard, A.E. Can. Inst. Food Sci.
    Technol. J. 1976, 9, 66.
2. Youngs, C.G. In "Oilseed and Pulse Crops in Western Canada -
    A Symposium"; Harapiak, J.T., Ed.; Western Co-operative
    Fertilizers Ltd., Calgary, 1975; Chap. 27.
3. Sosulski, F.W.; Youngs, C.G. J. Am. Oil Chem. Soc. 1979, 56,
    292.
4. Sosulski, F.W.; Elkowicz, K.; Reichert, R.D. J. Food Sci.
    1982, 47, 498.
5. Elkowicz, K.; Sosulski, F.W. J. Food Sci. 1982, 47, 1301.
6. Colonna, P.; Gallant, D.; Mercier, C. J. Food Sci. 1980, 45,
    1629.
7. Kon, S.; Sanshuck, D.W.; Jackson, R.; Huxsoll, C.C. J. Food
    Proc. Preserv. 1977, 1, 69.
8. Tyler, R.T.; Youngs, C.G.; Sosulski, F.W. Cereal Chem. 1981,
    58, 144.
9. Vose, J.R.; Basterrechea, M.J.; Gorin, P.A.J.; Finlayson,
    A.J.; Youngs, C.G. Cereal Chem.. 1976, 53, 928.
10. Lineback, D.R.; Ke, C.H. Cereal Chem. 1975, 52, 334.
11. Schoch, T.J.; Maywald, E.C. Cereal Chem. 1968, 54, 564.
12. Bramsnaes, F.; Olsen, H.S. J. Am. Oil Chem. Soc. 1979, 56,
    450.
13. Patel, K.M.; Johnson, J.A. Cereal Chem. 1974, 51, 693.
14. Sumner, A.K.; Nielsen, M.A.; Youngs, C.G. J. Food Sci. 1981,
    46. 364.
15. Vose, J.R. Cereal Chem. 1980, 57, 406.
16. Comer, F.W.; Fry, M.K. Cereal Chem. 1978, 55, 818.
17. Vose, J.R. Cereal Chem. 1977, 54, 1141.
18. Reichert, R.D. Cereal Chem. 1981, 58, 266.

19. Oomah, D.; Reichert, R.D.; Youngs, C.G. Cereal Chem. 1981, 58, 492.
20. Tyler, R.T.; Youngs, C.G.; Sosulski, F.W. Can. Inst. Food Sci. Technol. J. 1984, 17, 71.
21. "Approved Methods of the American Association of Cereal Chemists"; AACC, St. Paul, MN. 1983, 8th ed.
22. Lin, M.J.Y.; Humbert, E.S.; Sosulski, F.W. J. Food Sci. 1974, 39, 368.
23. Inklaar, P.A.; Fortuin, J. Food Technol. 1969, 23, 103.
24. Fan, T.Y.; Sosulski, F.W. Can. Inst. Food Sci. Technol. J. 1974, 7, 256.

RECEIVED January 24, 1986

16

Processing and Use of Dry, Edible, Bean Flours in Foods

Mark A. Uebersax and Mary E. Zabik

Department of Food Science and Human Nutrition, Michigan State University, East Lansing, MI 48824

Navy, pinto and black beans (Phaseolus vulgaris) were dry roasted in a solid-to-solid heat exchanger, dehulled by air aspiration, pin-milled and air-classified to yield whole, hulls, high protein, and high starch flour fractions. Proximate analyses, color, enzyme neutral detergent fiber (ENDF), nitrogen solubility index, oligosaccharide content, SDS-PAGE, and in-vitro digestibility were determined and resulted in differences due to bean types, mill fractions, and processing variables. Samples of all fractions were analyzed by emission spectroscopy (ICP) for minerals. Phytate phosphorus was present in the greatest quantity in the protein fraction (0.86-1.06%). Protein digestibility of the cotyledonary fractions, high protein and high starch, was similar for both bean types. However, digestibility of the hull fraction was greater for navy bean than that obtained from pinto bean. All flour fractions retained stability during conventional storage protocols. Quality attributes of foods incorporating fractions were objectively and subjectively evaluated. Acceptable cookies, donuts, quick breads and leavened doughs were produced using high fiber or high protein fractions at moderate levels of substitution (20%) for wheat flour.

Dry beans have been traditionally prepared by soaking and cooking in the home or consumed as commercially processed canned beans. Whole beans require soaking and cooking to ensure uniform expansion of the seed coat and hydration of the cotyledon matrix. Long cooking times required to achieve satisfactory palatability have impeded further utilization of dry beans. The use of dry edible beans could be readily expanded if they were available in the form of a shelf-stable flour.

Bean flours have been prepared by soaking and cooking beans in

0097-6156/86/0312-0190$06.00/0

water and steam followed by temperature dehydration over drum driers (1). Due to the high carbohydrate and high protein contents of legume seeds, research has been oriented to the production of high starch and high protein concentrates, using solvent extraction techniques either in pilot or commercial scale. The properties and functionality of these concentrates have been evaluated (2-5). Although this wet processing method results in high yields and high percentage values of protein and starch in the concentrates, it is energy intensive, generates large quantities of waste effluent, and results in a decline in nutritive value due to leaching.

Milling and fractionation of legumes to produce flour products have received increased interest in recent years. Milled products such as whole flour and air-classified high protein and high starch fractions from beans have demonstrated high nutrient retention and have provided increased versatility and improved utilization of beans. A number of studies on milling of legume flours and incorporation of flour into food products have been reported (6-9). Field peas have been successfully milled into whole flour and air-classified into high protein and high starch concentrates (6). Yield and compositional data of high protein and high starch concentrates of various legume seeds obtained through air-classification have been reported (7-9).

It was noted, however, that legume seeds contain antinutritional factors such as trypsin inhibitors and lectins which exert undesirable effects on nutritive value if they are not inactivated (10-11). In a recent study (12) navy beans were dry roasted in a particle-to-particle heat exchanger prior to milling and air-classification to maximize the nutritive potential of the whole, hulls, high protein, and high starch fractions.

The present research was conducted to evaluate selected physicochemical properties of the dry roasted and air-classified navy, pinto, and black bean flour fractions. Studies were conducted to determine the chemical composition and to characterize the functional properties of dry-roasted bean flour fractions and to evaluate the suitability of the flours for use in foods systems.

Dry Bean Processing

Dry beans (Phaseolus vulgaris), represented by the commercial classes of navy, pinto and black were used to produce flour fractions at the Food Protein Research and Development Center, Texas A & M University. Beans were dry-roasted under selected process conditions in a gas fired solid-to-solid heat exchanger, dehulled by air aspiration, pin-milled and air-classified to obtain four flour fractions. These fractions included whole, hulls, high protein, and high starch flours.

Roasting and Dehulling

The heat transfer medium of the exchanger consisted of 1.6 mm (1/16") diameter, type A, 90% aluminum oxide ceramic beads (Coors Ceramic Co., Golden City, CO) with a specific gravity of 3.6 g/cm^3. Selected roasting conditions were maintained by control of bead temperature and resident time. The beads were heated to 240°C and were maintained in the chamber with the raw beans for 100 seconds in a 1:5 ratio of beans to beads. These processing conditions resulted in an exit temperature of the beans of 113°C. Roasted beans were cracked through a corrugated roller mill (Ferrell

Ross; Oklahoma City, OK) into 6 to 8 pieces. The hulls, or seed coats, were removed using a zig-zag aspirator (Kice Metal Products; Wichita, KS). A hull flour fraction was produced by grinding the hull pieces through a swinging blade Model D6 Fitzmill (W.J. Fitzpatrick Co., Chicago, IL) using the impact surfaces for pulverization through a 0.69 mm (1/37") round hole screen.

Grinding and Air Classification

After hull removal, the cracked cotyledons were milled and air-classified at Alpine American Corporation pilot facilities; Natick, MA. The cracked cotyledons were finely ground in a Model 250 CW Stud Impact Mill at a speed of 11,789 rpm and a door speed of 5,647 rpm. The resulting flours were air-classified in a Model 410 MPVI Air Classifier at a rotor speed of 2,200 rpm and a brake ring setting of 3, using a 7.62 cm (3 inch) screw feeder operating at 25 rpm. Initially, two flour fractions were obtained: an intermediate starch fraction (coarse I) and an intermediate protein fraction (fines I). The intermediate protein fraction was reclassified under the following conditions: rotor speed of 2,200 rpm; brake ring setting of 0; 7.62 (3 inch) screw feeder operating at 25 rpm. As a result of this second air classification step, a high starch fraction (coarse II) and a high protein fraction (fines II) were obtained. The final materials produced as a result of the processes described included the hull flour fraction and three air-classified fractions: intermediate starch (starch I), high starch (starch II), and high protein (protein II). A flow diagram showing the flour processing scheme and the fractions produced appears in Figure 1. The percentage yield of each fraction of each bean type is shown in figure 2.

Physico-Chemical Characteristics

Composition

Chemical analyses of moisture, fat, dietary fiber, ash and protein were conducted on all flour fractions. Chemical measurements are important to characterize the specific nutritional properties of each fraction and to aid in determining each fraction's suitability for use in selected products. Figure 3 show the data from the average of eight treatments on chemical composition of roasted navy bean flour fractions. Moisture contents of these fractions ranged from 6.0 to 8.9%. Limited fat content was found in all frtions with the high protein fraction containing the greatest fat level (2.3%). Highest ash content was found in hull flour with slightly lower ash values in the remaining three flours. As expected, fiber content of hull flour was very high (40.5%) as was protein content (42.5%) of high protein fraction.

A Hunter Color and Color Difference Meter was used to evaluate flour color. The effect of roasting condition on Hunter L values of flours indicated that lightness values (L) decreased with increased roasting temperature and time with all navy bean cotyledonary fractions possessing a clear light appearance.

Identification and quantitative analysis of oligosaccharides by high performance liquid chromatography (HPLC) was done to determine potential problems with flatulence. Stachyose, which was highest in high protein flour, was the major oligosaccharide, followed by sucrose and glucose, with a trace amount of raffinose. Hull flour had the lowest total sugar level among fractions.

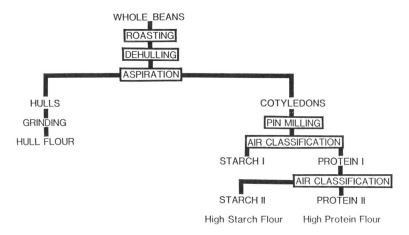

Figure 1.   Flow diagram of flour processing scheme. (Reproduced with permission from reference 33.)

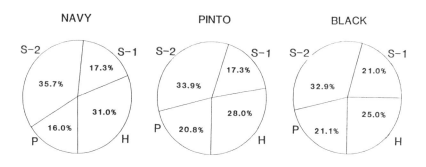

H – HULL   S–1 – STARCH I   S–2 – STARCH II   P – PROTEIN

Figure 2. Percentage yield of flour fractions obtained from navy, pinto, and black beans. (Reproduced with permission from reference 33.)

WHOLE NAVY BEAN FLOUR

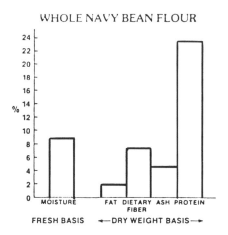

NAVY BEAN HULL FLOUR

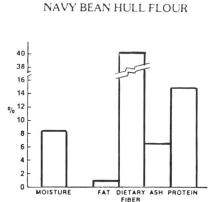

NAVY BEAN STARCH FRACTION

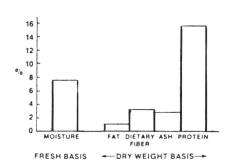

NAVY BEAN HIGH PROTEIN FRACTION

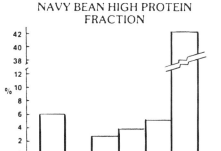

Figure 3. Proximate composition of dry roasted, air-classifed navy bean flour fractions.

Nitrogen Solubility Index (NSI) analyses were conducted on navy bean air-classified protein fractions to assess the influence of processing parameters on protein functionality. Nitrogen Solubility Index Method 46-23 (13) was modified by placing a 400 mL beaker containing the water-flour mixture on a mechanical stirrer and stirring at 120 rpm for 120 min at 30°C. Nitrogen solubility indices of protein flours are shown in Figure 4. Results indicated that increases in roasting temperature and time significantly reduced the NSI values.

Dough Mixing Characteristics

Three bean flour fractions were substituted for 10% bread flour to study rheological properties of the dough and the results are presented in Table 1. Farinograms showed changes in arrival time and a slight but consistent decrease in peak time for all fractions. Water absorption and dough stability were greatly affected. Substituting 10% bean hull flour had the greatest effect in increasing water absorption. The 60.8% water absorption value for wheat flour standard increased to 74.4% while the mixing stability changed from 6.0 min to 2.3 min. Whole bean flour had the least effect with a water absorption of 61.3% and a dough stability of 3.5 min. The effect of substituting the high protein fraction was slightly less detrimental than that which occurred with navy bean hulls. The increase in water absorption and reduction in dough stability due to the substitution of bean flour generally agreed with data reported by D'Appolonia (14) and Sathe et al. (15).

Phytic Acid and Mineral Content

Analyses were performed to determine the quantity and partitioning pattern of eight minerals and phytic acid among five flour fractions derived from navy, pinto, and black beans. After obtaining data on mineral content, phytic acid content, and proximate composition of each flour fraction of the three bean types, the study also attempted to correlate the individual microconstituents with macroconstituents. An inductively coupled plasma emission spectrometer was employed to measure the Zn, Fe, Ca, Cu, Na, K, Mg, and P content of each sample. The extraction and precipitation of phytic acid was accomplished according to the method of Wheeler and Ferrel (16). Makower's (17) method was then used to convert ferric phytate to ferric hydroxide. The AOAC (18) method for measuring ferric iron using ortho-phenanthroline was followed. Phytate phosphorus content was calculated from iron concentration by assuming a 4:6 atomic ratio of iron to phosphorus (16). Phytic acid content was estimated on the basis that 28.20% of the weight of phytic acid is contributed by.

Overall, the large number of significant differences among the fractions of all three bean types, reveals that the various minerals studied were not equally distributed among the fractions, except for sodium, but rather that partitioning occurred. Consistent with the ash values obtained, the protein flour fractions of the three bean types contained larger amounts of Fe, Mg, P, Z, and K whereas the starch II fractions contained smaller amounts of these minerals and Ca and Cu than the other flour fractions. Phytic acid content ranged from 8.7-30.2 mg/g for navy flours, from 4.3-23.7 mg/g for pinto flours. Total phosphorus content correlated well with both phytic acid content and protein content. Phytic acid content was

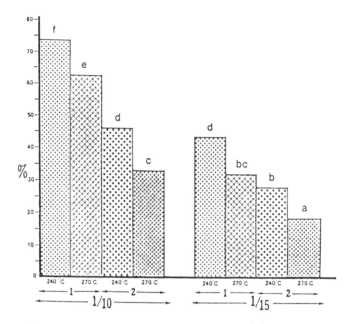

Figure 4. Mean nitrogen solubility index (%) for protein flour dry roasted under various processing conditions: bean/bean ration (1/10, 1/15); residence time (1, 2 min); and bead temperature (240, 270 C). Means with similar letters are not signicantly different, P<0.05). (Reproduced with permission from reference 34.)

also highly postivitely correlated to the protein content. The iron,
Zn, P and Mg contents of the navy, pinto, and black bean flours
showed positive correlations with both protein content and and
phytic acid content, which suggest that these minerals may become
bound by phytic acid as possibly by the protein. Calcium did not
correlate with protein content, phytic acid content, or starch
content. The hull flour fractions of navy and black beans had both
the highest dietary fiber contents and the highest calcium contents
of all fractions studied. This suggests that calcium may be bound
by fiber. As such, any processing techniques implemented to isolate
protein may also serve to increase the concentration of phytic
acid. In addition, the high degree of correlation achieved between
protein content and the contents of Zn, Fe, K, and Mg, and between
phytic acid content and these minerals suggests that these elements
are present as metallic phytates.

In Vitro Protein Digestibility

The in vitro digestibility of the air classified fractions of navy
and pinto beans was assessed by measuring the extent to which the pH
of the protein suspension dropped when treated with a multi-enzyme
system which included trypsin, chymotrypsin, and peptidase as
described by Hsu et al. (19) and modified by Satterlee et al. (20)
to include a fourth enzyme, the protease of Streptomyces griseus.
Values were calculated for Phaseolus vulgaris according to Wolzak et
al. (21). Results are presented in figure 5. Roasting increased
protein digestibility for each bean type and for all flour
fractions. Protein digestibility of high starch and high protein
fractions was not significantly different for either bean type.
Hull fractions possessed less protein digestibility than did
cotyledon fractions. Raw pinto hull protein was less digestible
than navy hull protein. However, heating resulted in improvement to
yield non-significant differences between navy and pinto hulls. The
SDS-PAGE method of Weber and Osborne (22) was followed to study the
distribution of bean proteins among the different air classified
fractions. SDS-PAGE demonstrated roasting resulted in a shift to
broader and higher molecular weight bands with globulins exhibiting
greater heat stability than albumins.

Storage Stability

Equilibrium Moisture Content (EMC)

The effect of roasting conditions on the EMC of the air-classified
navy bean flour fractions in various water activity levels was
determined at room temperature. Figure 6 illustrates the water
asorption isotherms of bean flour (Fig. 6) fractions for different
roasting treatments at various a_w levels after eight weeks of
storage. Roasting treatment did not affect the EMC of the flours to
a significant extent for the two lowest a_w levels (0.48 and 0.53);
however, EMC of whole flours decreased with increased roasting time
and temperature and the effect of time/temperature treatment was
shown to be significant with increased a_w, especially with the a_w's
beyond 0.8.

A similar roasting effect was shown for the water adsorption
isotherms of hull flours except that the most severe roasting did
not affect the EMC as dramatically as it did whole flours for the
highest a_w. The water asorption isotherms of high protein flours
indicate that roasting did not affect EMC of these flours

Table 1. Farinograph values of dry-roasted navy bean flour fractions[1,2].

Paramater	Flour Fractions			
	Whole Flour	Hull Flour	High Protein Flour	Wheat Flour Standard
Curve Type				
Water Absorption (%)	61.3±0.5 [a]	74.4±0.1 [c]	71.1±0.1 [b]	60.8±0.2 [a]
Arrival Time (min.)	3.0±0.1 [a]	3.5±0.0 [c]	3.2±0.0 [b]	3.0±0.1 [a]
Peak Time (min.)	4.3±0.1 [a]	4.4±0.1 [a]	4.3±0.1 [a]	5.0±0.1 [b]
Stability (min.)	3.5±0.1 [c]	2.3±0.3 [a]	3.0±0.1 [b]	6.0±0.2 [d]

[1]Mean values and standard deviations (like letters within each row indicate no significant differences at P < 0.05 by Tukey mean separation; n = 3)

[2]Bean fractions substituted 10% for wheat flour

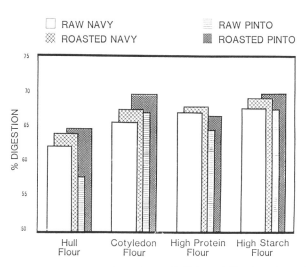

□ RAW NAVY   ☰ RAW PINTO
▨ ROASTED NAVY   ■ ROASTED PINTO

Figure 5. In vitro protein digestibility (%) for fractions of raw and roasted navy and pinto beans.

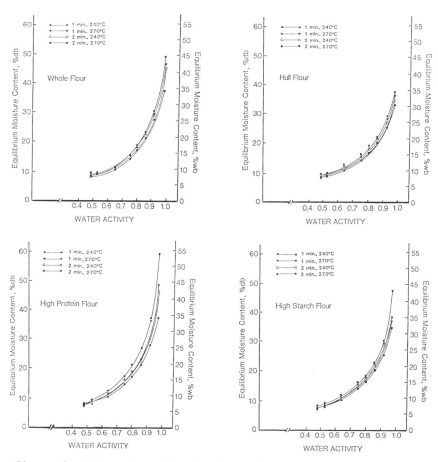

Figure 6. Water adsorption isotherms by dry-roasted navy bean flour fractions at room temperature (20°C).

significantly for the lowest a_w level (0.48) and similar isotherms were obtained for flours from roasting treatments of 1 min/270°C and 2 min/240°C. In general, EMC of high protein flours decreased significantly with increased roasting time and temperature over the range of a_w in which flours were stored. Isotherm curves of high starch flours as affected by roasting were similar to those for whole flours except that slightly lower EMC values were found for high starch flours at higher a_w levels. Comparisons of isotherms from this figure demonstrated that high protein flours having the highest protein content (41.3% to 47.6% db) had the highest EMC among all fractions, while hull flour with somewhat lower protein content (14.1% to 18.4%) had the lowest EMC at all a_w levels. Generally, moisture content of samples beyond 16% db caused significant increases of mold growth for all the flour fractions. High protein flour had the highest water uptake and thus resulted in the most mold growth, whereas the high starch fraction absorbed less water and had the least mold growth. EMC of the hull flour after eight weeks was lower than 40% db; however, there was considerable mold growth on the flour surface.

Property Changes During Storage

Following storage (3 months) the samples at a_w higher than 0.75 were moldy and thus discarded. Since the roasting treatment of 1 min at 240°C was demonstrated to obtain the highest EMC after storage, flours from this treatment were selected for color, NSI, and sugar analysis. Hunter L values remained unchanged with increased a_w for all flour fractions; however, a_L and b_L values of these fractions increased with increased a_w indicating that flours became more red and more yellow toward high a_w levels. Labuza et al. (23) stated that non-enzymatic browning increased with increased a_w reaching a maximum and then decreased due to the dilution of the reactants. The fact that discoloration of flour increased with increased a_w in this study could be due to the increased browning reactions in the stored samples at high a_w levels. NSI values were found to decrease significantly for whole flours and high starch flours stored beyond 0.48 a_w and for high protein flour stored beyond 0.53 a_w. In general, NSI values of these fractions decreased with increased storage a_w except for hull flour. NSI values of hull flours were the lowest among fractions and did not show any significant change due to increased a_w levels. Sugar contents of flours were analyzed by HPLC and, in general, glucose decreased significantly with increased a_w for all fractions. Other sugars remained unchanged or decreased with increasing a_w. The reduction of glucose was probably attributed to the increased non-enzymatic browning reactions which also caused the increased discoloration of flours at higher a_w levels.

To further evaluate the stability of flours during long term storage, flour fractions from the optimum roasting treatment were selected and stored at two temperatures (20°C, 37.7°C) and three preadjusted moisture contents (6%, 9%, 12% db). It was concluded that flours of all fractions stored at less than 0.75 a_w for two months at ambient temperature resulted in stable quality. Long term storage indicated that the flours stored up to 24 months with 6% and 9% moisture at 20°C or with 6% moisture at 37.7°C retained good quality.

Bean Product Utilization in Wheat Flour Based Foods

The utilization of bean flour of varying extractions from untreated and dry-roasted split navy beans has been investigated for bread making (14). Functional properties of starch flour prepared from various legumes were studied by Schoch and Maywald (2) and Naivikul and D'Appolonia (24). Rheological properties of dough and baking quality of bread as affected by the addition of navy bean flour and protein concentrate were studied (15). Cereal brans have been incorporated into bread, cakes, and cookies to produce baked products with high fiber contents (25-39). Research conducted by DeFouw et al. (29) concluded that navy bean hulls were an acceptable source of dietary fiber in spice-flavored layer cakes.

Cake Donuts. The effects of 0-30% substitution of dry-roasted air-classified navy, pinto or black protein flour for wheat flour on the physical and sensory qualities of cake donuts were investigated. In general, fat absorption decreased as the level of substitution increased with the pinto bean variable having the lowest values. An increase of navy bean and pinto protein levels decreased donut height after the 10% level. The donuts with pinto bean protein exhibited less spread than the other bean type donut variables. Navy bean protein produced a more tender donut than the other bean types. The control donut and those with navy bean protein were the lightest in color, followed by pinto and black bean protein donuts, respectively. With an increase in protein level incorporation over 10%, a darker donut was produced. The navy and pinto protein flour donuts were brownish, whereas the black protein flour produced a grayer donut. The sensory panel found the donuts containing black bean protein to have poor color, but all of the other variables had an acceptable color.

Consumer panels indicated that substitution of 13% of the dry mix (20% of the flour) with navy or pinto protein was highly acceptable. These donuts also had less fat, were softer and exhibited less firming after six days of storage.

Sugar Snap Cookies. Quality characteristics of sugar snap cookies incorporating dry bean hulls were evaluated in a factorial experiment. Navy bean hulls were obtained from three roasting treatments: none, moderate temperature and high temperature roasting. Cookies were prepared substituting 0, 10, 20 and 30% hulls for flour. Cookie qualities significantly affected by substitution level included: surface characteristics, color, moisture, tenderness, and flavor. Major differences attributed to roasting were color, moisture and surface characteristics. Flavor was not significantly different between cookies prepared with either roasted or unroasted navy bean hulls.

Spiced Layer Cake. Investigation was undertaken to evaluate navy bean hulls as an alternative source of dietary fiber in spice-flavored layer cakes and to compare hull flour from beans with no heat treatment to hull flour from beans roasted for 2 min at 240°C. Results of the objective analyses performed on both the batter and the cakes indicated that the addition of 15% navy bean hulls resulted in a thicker batter, due to the high water absorbancy of the hulls; however, only the batter containing the unroasted hulls was significantly more viscous than the control batter. Cakes prepared with the roasted navy bean hulls tended to be slightly more moist and tender than the control, but these results were not significant. Shrinkage was greater for cakes prepared with the unheated hulls than for cakes prepared with roasted hulls, although neither differed significantly from the control. The addition of navy bean hulls darkened the crumb color of cakes. Cakes prepared with the unroasted hulls were significantly darker than either the control or cakes made with roasted hulls; cakes prepared with roasted hulls were significantly darker than the control. The unheated hulls were slightly gray and imparted this darkness to cakes. The roasted hulls were yellow; consequently, cakes prepared with them were the most yellow; however, differences were not significant. Redness of the cakes did not differ significantly because of hull substitution.

Sensory analyses indicated that flavor and general acceptability of the cakes prepared with unheated bean hulls were significantly less desirable than the control; however, the were still rated as moderately desirable. For all sensory characteristics, cakes prepared with roasted hulls scored higher than cakes prepared with unheated hulls and were closer to values given to the control cakes.

Both the unheated and roasted navy bean hulls are high in dietary fiber (43% and 40%, respectively), based on ENDF determinations.

Few differences were attributed to roasting of the navy bean hulls; however, color of the hulls improved with heat treatment. Cakes prepared with either unheated or roasted bean hulls compared favorably to the control. Consequently, navy bean hulls are an acceptable source of dietary fiber in a flavored cake. More research should be initiated to investigate alternative uses for navy bean hulls as a dietary fiber source.

Pumpkin Quick Bread. Previous research yielded legume breads and bread products with decreased flavor, volume, and color values (14,30-31). The present research attempted to overcome these difficulties by producing a quick bread of high flavor intensity through the use of spices, consistent specific volume through the method of preparation, and decreased color change through the use of an already darkly colored product. A study was designed to compare the quality characteristics of pumpkin bread containing 0, 20, 35 and 50% whole navy bean flour substituted for wheat flour. Batter viscosity, specific gravity, and pH for all levels of navy bean incorporation were similar. Color of the bread was darker (decreased L value), less red (decreased a_L value), and less yellow (decreased b_L value) with increasing levels of navy bean flour substitution. Although significant differences in yellowness and redness occurred for each level of navy bean flour substitution, lightness values different only between the control and the pumpkin bread substituted with 50% navy bean flour. Moisture content of the baked products, however, did not differ significantly. Volume of the quick breads decreased with increasing levels of navy bean flour substitution, with differences most significant between the control and bread with 50% navy bean flour. Pumpkin bread tenderness increased with increasing amounts of bean flour substitution; however, no statistical differences were found.

Odor, color, and tenderness of the pumpkin breads were not affected by navy bean flour substitution. Texture scores for pumpkin bread with the 35% level of navy bean flour substitution were superior to those of the control. Moistness of the bread with 50% navy bean flour, however, was significantly lower than the other bread variables. Flavor scores decreased increasing levels of substitution; however, a statistical difference was noted only between the control and the bread with the 50% level of navy bean flour. Moreover, all pumpkin breads were of good quality, scoring above five on a seven-point scale; however, the pumpkin bread containing 35% navy bean protein substitution appeared to be optimum.

The legumes have a high lysine content (7,32), which makes them an excellent complement to cereal proteins. Protein contents of pumpkin breads were 5.0, 5.6, 6.2 and 6.7% for loaves that had 0, 20, 35 and 50% Navy bean flour, respectively.

Quick breads are unique baked products in that they encompass some of the functional qualities of both bread and cake. The products are baked as breads, yet they lack a strong gluten development and contain a high ratio of sugar to flour, as in a cake system. Thus, incorporation of protein substitutes is feasible. The results of the study agreed with findings of other researchers. Tenderness values increased because of reduced gluten formation and the diluted nature of the gluten. Volume decreased because of decreased gluten formation. The product was darker because of increased reducing sugars in the navy bean flour, promoting Maillard browning. Nevertheless, a high-quality pumpkin bread was produced with navy bean flour substituted for 35% of the flour. This bread contained about 25% more protein than the control.

Peanut Butter Cookies. Three types of peanut butter cookies, control, high protein (30% bean protein concentrate substitute for flour) and high fiber (20% bean hulls substituted for flour) prepared at Michigan State University bakery were evaluated for their physical characteristics (Table 2).

Table II. Quality characteristics of peanut butter cookies prepared with navy bean hull (20%) and protein (30%) substitutions for wheat flour.

Treatment	MOISTURE (%)	BAKING LOSS (%)	SPEED FACTOR (W/T)	TENDERNESS (LB/G)	SENSORY SCORE HIGHEST 7 = RATE
CONTROL	5.84	10.73	10.0	25.73	5.9
BEAN HULL (20%)	5.03	10.45	9.6	31.58	5.9
BEAN PROTEIN (30%)	7.04	11.23	9.1	21.62	5.5

N = 3

High protein cookies had the highest moisture content and baking loss, but lowest spread factor. Tenderness of high fiber cookies was shown to be the highest while that of high protein cookies exhibited the lowest tenderness value. These cookies were also evaluated using a seven-point hedonic scale by 300 participants at (Focus:HOPE, Detroit, Michigan). These panelists included wide representation of children, teenagers and adults. Black was the predominant race although some whites and other races took part. Panelists' general comments were favorable. Data showed the high fiber cookies to score as high as the control while the high protein cookie scored slightly lower. Even though statistical differences occurred, all cookies were rated acceptable by these panelists.

CONCLUSIONS

Dry bean flour fractions produced by dry roasting, milling and air classification resulted in versatile food ingredients. Fractions possessed good functional and nutritional properties which were found to be acceptable in a variety of food systems. These processes and products appear to have potential for improving nutritive status through improved dry bean utilization.

ACKNOWLEDGMENTS

This research was conducted with USDA funding. The authors wish to express appreciation to Dr. E. Lusas and coworkers at the Food Protein Research and Development Center, Texas A & M University for their cooperative research efforts in process and product development.

LITERATURE CITED

1. Bakker-Arkema, F.W.; Patterson, R.J.; Bedford, C.L. Trans. ASAE 1967, 12, 13.
2. Schoch, T.J.; Maywald, E.C. Cereal Chem. 1968, 45, 564.
3. Sathe, S.K.; Salunkhe, D.K. J. Food Sci. 1981, 46, 71.
4. Chang, K.C.; Satterlee, L.D. J. Food Sci. 1979, 44, 1589.
5. Vose, J.R. Cereal Chem. 1980, 57, 406.
6. Vose, J.R.; Basterrechea, M.J.; Gorin, P.A.J.; Finlayson, A.J.; Youngs, C.G. Cereal Chem. 1976, 53, 928.
7. Patel, K.M.; Bedford, C.L.; Youngs, C.W. Cereal Chem. 1980, 57, 123.
8. Tyler, R.T.; Youngs, C.G.; Sosulski Cereal Chem. 1981, 58, 144.
9. Sahasrabudhe, M.T.; Quinn, J.R.; Paton, D.; Youngs, C.G.; Skura, B.J. J. Food Sci. 1981, 46, 1079.
10. Puztai, A. Nutr. Abstr. Rev. 1967, 37, 1.
11. Jaffe, W.G. In "Nutritional Improvement of Food Legumes by Breeding"; John, Wiley and Sons: New York, 1975.
12. Aguilera, J.M.; Lusas, E.W.; Uebersax, M.A.; Zabik, M.E. J. Food Sci. 1982, 47, 1151.
13. "Approved Methods", Am. Assoc. Cereal Chemists (AACC), 1962, 7th ed.
14. D'Appolonia, B.L. Cereal Chem. 1978, 55, 898.
15. Sathe, S.K.; Ponte, J.G. Jr.; Rangnekar, P.D.; Salunkhe, D.K. Cereal Chem. 1981, 58, 97.
16. Wheeler, E.L.; Ferrel, R.E. Cereal Chem. 1971, 48, 313.
17. Makower, R.U. Cereal Chem. 1970, 47, 288.
18. "Official Methods of Analysis", Assoc. Offic. Agric. Chemsits (AOAC), 1980, 13th ed.

19. Hsu, H.W.; Vavak, D.L.; Satterlee, L.D.; Miller, G.A. J. Food Sci. 1977, 42, 1269.
20. Satterlee, L.D.; Marshall, H.F.; Tennyson, J.M. J. Am. Oil Chem. Soc. 1979, 56, 103.
21. Wolzak, A.; Bressani, R.; Gomez Brenes, R. Qual. Plant. Plan Foods Hum. Nutr. 1981, 31, 31.
22. Weber, K.; Osborn, M. J. Biol. Chem. 1969, 244, 4406.
23. Labuza, T.P.; Tannenbaum, S.R.; Karel, M. Food Technol. 1970, 24, 35.
24. Naivikul, O.; D'Appolonia, B.L. Cereal Chem. 1979, 56, 24.
25. Pomeranz, Y.; Shogren, M.D.; Finney, K.F.; Bechted, D.B. Cereal Chem. 1977, 54, 25.
26. Rajchel, C.L.; Zabik, M.E.; Evenson, E. Bakers Digest 1975, 49, 27.
27. Shafer, M.A.M.; Zabik, M.E. J. Food Sci. 1978, 43, 375.
28. Vratanina, D.L.; Zabik, M.E. J. Food Sci. 1978, 43, 1590.
29. Defouw, C.; Zabik, M.E.; Uebersax, M.A.; Aguilera, J.M.; Lusas, E. Cereal Chem. 1982, 59, 229.
30. Guadagni, D.G.; Venstrom, D. J. Food Sci. 1972, 37, 774.
31. Satterlee, L.D.; Bembers, M.; Kendrick, J.G. J. Food Sci. 1975, 40, 81.
32. Bolourforooshan, M.; Markakis, P. J. Food Sci. 1979, 44, 390.
33. Tecklenburg, E.; Zabik, M. E.; Uebersax, M. A.; Dietz, J. C.; Lusas, E. W. J. Food Sci. 1984, 49, 570.
34. Zabik, M. E.; Uebersax, M. A.; Lee, J. P.; Aguilera, J. M.; Lusas, E. W. J. Am. Oil Chem. Soc. 1983, 60(7), 1305.

RECEIVED December 26, 1985

17

Winged Bean as a Source of Protein: Recent Advances

Sachi. Sri Kantha and John W. Erdman, Jr.

Department of Food Science, University of Illinois, Urbana, IL 61801

Protein quality studies evaluating various edible portions of the winged bean plant and the results of recent International Field Trials are reviewed. Research
efforts have been focused upon the mature seed (protein
content 20.7-45.9% in 240 accessions), which is occasionally eaten in parts of Indonesia and Papua, New
Guinea. Autoclaved seed meal and wet heat treated seed
meal provided a corrected PER value of 1.76 and 1.72
respectively, in comparison to casein control of 2.50.
Of the 190 accessions evaluated after one year's growth
in Florida, 38 showed tuber formation. Tender leaves
are incorporated in the food preparations in the South
Asian region. Leaf protein concentrate (crude protein
51.9%) prepared from leaves yielded a PER value of 2.2
in comparison to 2.7 for corn-soy control. International Winged Bean Trials conducted in 19 countries recommend
5 varieties as having the best yield potential under
varying environmental conditions.

The winged bean (Psophocarpus tetragonolobus L.DC) is a twining,
perennial herbaceous legume which can climb to 3-4 meters. Botani-
cally it is classified as belonging to Family Leguminoseae, sub-
family Paplionoideae, tribe Phaseoleae and subtribe Phaseoline
(1,2). Of the 9 recognized species of Psophocarpus, four are
presently cultivated as a home garden crop in the humid tropics of
South and South East Asia. Three species (P.scandens, P.palustris
and P.grandiflorus) are now being cultivated on experimental basis
as forage and cover crops (3). Though it has yet to attain the
status of a commercially productive legume, the nutritional aspects
of winged bean have been studied in USA, India, Japan, Malaysia,
Philippines, Papua New Guinea, Czechoslovakia, Australia and Sri
Lanka.
    The actively investigated areas and the locations where winged
bean studies have been conducted are shown in Table I. Since 1975,
on a conservative estimate, approximately 150 research reports
covering areas such as microstructure of winged bean protein, nu-

0097-6156/86/0312-0206$06.00/0
© 1986 American Chemical Society

Table I

Studies on the nutritional aspects of winged bean

Research area	Location	References of reported studies
1. Microstructure of winged bean protein	Australia Japan	(4-7) (8-10)
2. Nutrient, anti-nutrient composition and isolation		
(a) Nutrient analysis, bioavailability studies, functional properties of protein	USA Sri Lanka India Philippines	(11-16) (17) (18,19) (20)
(b) Protease inhibitors, tannins	USA Sri Lanka Australia Malaysia	(21-24) (25) (26-30) (31,32)
(c) Phytohgaemagglutinins(lectins)	USA Sri Lanka Japan India	(33) (34) (35-37) (38)
3. Storage studies on seeds	Japan	(39)

(Continued on next page)

Table I (cont.)

Studies on the nutritional aspects of winged bean

Research area	Location	References of reported studies
4. Nutritive value in animals		
(a) rats	India	(18,40)
	Czechoslovakia	(41)
	Japan	(42)
	Peru	(43)
(b) chicken	Malaysia	(44)
	USA	(45)
5. Human experiments and trials	Czechoslovakia	(46,47)
6. Preparation of foods		
(a) bread	USA	(48-50)
(b) tempeh	Indonesia	(51)
(c) miso	Japan	(52)
(d) tofu	Japan	(53)
	USA	(54,55)
7. Consumer acceptability studies	Sri Lanka	(56,57)
	Papua New Guinea	(58)

trient and anti-nutrient composition, effect of storage on seed
components, protein quality estimation of seed flour for chickens
and rats, human feeding trials, preparation of foods from processed
winged bean flour and consumer acceptability studies have been published. A representative list of about one third of the publications (4-58) are noted in Table I.

Previously published reviews (59-61) have analyzed in detail
the emergence of winged bean as a legume of significant potential.
This paper summarizes recent reports on protein quality evaluation
of the edible portions of this legume.

Seeds

Nutrition research on winged bean has mainly focused upon the
tough, mature seed, which is rich in protein and oil. The seeds
are occasionally consumed in Indonesia and Papua New Guinea (59,
60). A survey of 240 winged bean accessions from 16 countries
showed that the protein and oil contents range between 20.7-45.9%
and 7.2-21.5% respectively (13). Studies on the evaluation of seed
flour as an alternative feed source for livestock has only recently
begun. The two most commonly used plant protein sources in broiler
(chicken) diets to date have been soybean and peanut cake (62).
The effect of replacing soybean meal with winged bean meal at 0,
19, 44, 74 and 95% on a protein basis was studied by de Lumen et
al. (45). No statistically significant differences were observed
among different rations in metabolizable energy, nitrogen retention
and broiler performance. However, there was an indication of adverse effects at 95% replacement. In another experiment using more
birds per treatment, the same researchers reported that replacement
at 75% and 100% level decreased the metabolizable energy and led to
poor broiler performance. Though defatted winged bean meal prepared by autoclaving at 121°C for 30 min was used in this study, it
was inferred that hulls in the seeds may have caused the decrease
in metabolizable energy that led to the poor performance of broilers.

Two studies (44,63) had shown that growth performance and nitrogen retention efficiency of chicks were severely reduced by the
inclusion of raw full fat winged bean seeds at rates of up to
400g/kg in semi-purified diets. Heat treatment of winged beans
prior to dietary incorporation results in reversal of the growth
depression and increases the efficiency of food utilization.

In a recent investigation performed in Nigeria (62), farm processed winged bean was fed to broilers in partial or complete replacement for either soybean meal or peanut cake. The feed gain
ratio obtained for birds fed on winged bean, soybean and peanut
cake diets were 3.0, 2.8 and 2.7 respectively. However, the study
revealed that mortality of broilers fed on winged bean-based diet
was high (14%), suggesting that the bean cake was not effectively
processed to remove the anti-nutrients. Results of another study
(64) showed that it was feasible to feed autoclaved winged beans to
Japanese quail chicks. However, the authors suggest that prior to
feeding quail or other monogastric animals, the type of heat processing treatment must be carefully monitored. Beans heated for 30

min at 118°C produced excellent growth response. The addition of
0.8% methionine to winged bean meal was suggested to be necessary
for equivalent growth compared with an addition level of 0.3% in
soybean meal diets (64).
These investigations show the potential value of full fat
winged bean as a broiler feed. However, further studies that de-
fine conditions for destruction of anti-nutrients which cause reduced growth performance should be performed.
Quite a number of studies (18,40-43,65,66) published in the
last few years have investigated the growth performance of
laboratory rats fed raw or autoclaved winged been. In a general
context, rat studies provide a good model to evaluate adequate
processing techniques needed for elimination of anti-nutrients.
There is uniformity among the reported studies regarding the lethal
effect of feeding raw winged beans to rats. While raw bean meal
results in poor or no growth, autoclaved meal results in adequate
growth and no toxic side effects in the rat.
Table II provides a comparison of four protein efficiency ratio (PER) studies of heat-processed winged bean seed flour. In
these studies rats were fed 10% protein level of casein on test
protein for 28 days. PER was determined by evaluating grams growth
per grams protein intake (times 100). The seeds used in the four
studies were subjected to varying time and temperature heat treatments. However it can be seen that the PER values obtained for
winged bean compare favorably with that of casein control. In a
study from our laboratory (66) it was found that dehulling followed
by moist heat treatment for 30 min could be utilized as a good procedure of heat treatment in place of autoclaving. If these data
are confirmed with human trials, it could enhance the use of seeds
since in the tropical developing countries where other portions of
winged bean are already consumed and facilities for autoclaving are
unavailable for rural peasants.
Tubers: Tubers are quite commonly consumed by villagers in Burma.
Other tropical countries also report sporadic use of winged bean
tubers (59). However, in comparison to the available data on the
nutrient, anti-nutrient composition and the protein quality of the
seed flour, published data on the nutritional value of winged bean
tuber is somewhat limited. The few reported studies dealing with a
larger number of varieties grown in different locations (25,67-71)
show that tubers are mainly composed of protein and carbohydrates.
Fresh tubers contain a moisture content of 55.5-65.8% and a
rather high crude protein content of 5.5-5.6%. Air-drying or
freeze drying of tubers results in an increase of protein content
to 13.1-16.5% (Table III). However, dry weight of tuber per plant
was low, varying from less than 25g to about 50g in different winged bean cultivars (72). Furthermore, of the 190 accessions evaluated after one year's growth in Florida, only 38 showed tuber formation (72).
According to Masefield (73) the winged bean nodulated well
wherever the crop has been grown, irrespective of whether the seed
was inoculated or not. The mean protein contents (dry wt. basis +
S.D.) of inoculated and uninoculated tubers of six strains originating from Sri Lanka, Nigeria, Indonesia, Papua, New Guinea and
Thailand are reported as 20.0(+3.7) and 17.3(+6.2), respectively

Table II

Comparison of PER studies on heat-processed winged bean seeds
in rats (fed at 10% protein level)

Sample description and preparation for 10% protein diet	Reported PER values		Corrected PER values		Reference
	casein	winged bean	casein	winged bean	
Ghana: soaked, dehulled cotyledon, antoclaved 100°C for 30 min.	3.04	2.14	2.50	1.76	(41)
Papua New Guinea: antoclaved heterogeneous sample, 118°C for 20 min.	3.00	1.60	2.50	1.33	(65)
Peru: soaked seeds, boiled for 60 min.	3.09	1.73	2.50	1.44	(43)
Sri Lanka: soaked, dehulled cotyledon, boiled for 30 min.	3.16	2.17	2.50	1.72	(66)

Table III

Chemical composition of winged bean tubers grown in different countries
(Mean values)

	Fresh Tubers		Dried Tubers	
	Thailand (3 varieties)	Sri Lanka[b] (19 varieties)	Philippines[c] (26 varieties)	India[d] (12 varieties)
Moisture (%)	55.5	65.8	5.3	9.6
crude protein (%)	5.5	5.6	16.5	13.1
crude fat (%)	-[f]	-	1.0	-
ash (%)	-	-	2.9	2.0
fiber (%)	-	-	16.6[g]	2.7
carbohydrate (%)[e]	-	-	57.7	70.1
starch	26.1	-	-	70.1
reducing sugars (mg/100g)	-	-	-	108.6
non-reducing sugars (mg/100g)	-	-	-	71.0

a. ref. (69)
b. ref. (25)
c. ref. (67); freeze-dried sample
d. ref. (68); air-dried sample

e. calculated by difference
f. not determined
g. calculated by difference

(<u>71</u>). There was no significant difference between the amino acid
contents of the inoculated or uninoculated tubers.
   The Rhizobial characteristics of the winged bean nodules were
reported by Ikram and Broughton in a series of studies (<u>74-76</u>). Of
the Rhizobia isolated from 14 different genera of legumes when
tested for their nitrogen fixing effectiveness with <u>P</u>. tetragonolobus, all but one isolate was able to form nodules with winged
bean. Thus, Ikram and Broughton (<u>74</u>) suggested that the winged
bean may be considered as promiscuous with respect to its Rhizobial
requirements.
   From a study of 6 varieties of tubers from different regions
of Asia and grown at two locations (Western Australia and
Malaysia), Kortt and Caldwell (<u>70</u>) concluded that,
   1. the reported variations in the composition of tuber crude pro-
      tein are probably due to variations in the content of free
      amino acids and low molecular weight peptides accumulated by
      the tuber under different nutrient growth conditions.
   2. the development of the tuber as a food crop is dependent more
      on agronomic factors such as variety selections, yields, ease
      of harvest and storage of tubers, than on nutritional limi-
      tations.
Well designed studies on nutritional quality of tuber protein with
humans, poultry or other farm animals have yet to be reported.
<u>Leaves</u>: Among the variety of green, leafy foods, winged bean
leaves are one of the richest sources of protein and carotenoids
(<u>59</u>). In Sri Lanka and elsewhere, they are consumed after steaming
or by direct incorporation into salads (<u>60</u>). Leaf protein concentrates (LPC) prepared from winged bean leaves grown at the experimental plots in Puerto Rico were reported to contain 51.9% crude
protein and an essential amino acid profile similar to the reported
values of alfalfa LPC. Winged bean LPC was rich in leucine, lysine, phenylalanine and valine. The only growth performance study
conducted in rats yielded a PER value of 2.2 for winged bean leaves
in comparison to 2.7 for a corn-soy control (<u>77</u>). Clearly more
studies are needed on leaf protein.
<u>Haulm</u> (stem): The possibility of using winged bean haulm as a po-
tential raw material for single cell protein production in the
tropics has been investigated by Zomer et al. (<u>78</u>). Semi-solid
fermentation of haulm, using various fungi, yielded a product containing 20% true protein, a 30% increase over the native material.
The fermentation also increased the sulfur amino acid content by
1.5 to 6 fold.

International Winged Bean Trials

After the First International Winged Bean Symposium held at Los
Banos, Philippines in 1978, the first international winged bean
trials were launched. According to the report of Khan (<u>79</u>), who
served as the coordinator of the trials, the aims of the trials
were to:
   i) test adaptation of selected winged bean varieties over a wide
      range of environmental conditions;
  ii) provide sources of germ plasm of a new crop which cooperators
      may use directly;
 iii) identify areas of the world which have potential for winged
      bean production; and

iv) test the response of winged bean to different environments. Altogether 31 researchers from 19 countries participated in the trials in regions as diverse as Bangladesh, Indonesia, Korea, Taiwan, Ivory Coast, Guatemala and USA. Through most of the trials sites were located in tropical climatic regions, a few were held in the subtropics as well with three additional regions (Australia, Korea and Pakistan) at latitudes 31°-37°. Of the 14 varieties used in the trial, five varieties were selected to have the best potential. Table IV shows the details obtained for these 5 varieties and the recommendations made by the evaluators. The Thailand variety (Nakhan Sawan) produced the highest yield in seeds and dry pods. Indonesian varieties (LBNC1 and LBNC3) produced higher yields of fresh tubers. However, these three Southeast Asian varieties were late maturing types and were also so sensitive to long day lengths that in some trials flowering did not occur. Two Papua, New Guinean varieties (UPS 121 and UPS 122) were identified as having medium vigor. Though winged bean genotypes from various regions of Southeast Asia exhibit considerable genetic diversity, Kortt (30) reported no differences in protein subunit composition. Thus it seems at this juncture, that variation in protein subunit composition cannot be exploited by plant breeders to improve the nutritional value of the seed. This area needs further confirmatory research.

Future areas of research

The winged bean was able to rise from comparative obscurity to some degree of prominence in the past decade, mainly because it was promoted as a legume in which all portions of the plant were edible and of high nutritional value. Some caution must be exercised at this point since all portions cannot be exploited from the same plant (80). Pod and seed production is influenced by plucking the leaves and/or flowers. Also, high tuber-yielding varieties do not produce high seed yields. Furthermore, a number of problems must be overcome before the winged bean becomes an important cash crop for the tropics.

Presently, the expansion of winged bean production and commercial development is inhibited by the lack of dwarf morphological variants and ecological uncertainties (80). A preliminary note on the development of bush type mutant in India has appeared (81), though scientific details have not been reported yet.

Another area which needs additional research is the nature of photoperiod sensitivity of the plant. Most, if not all, winged bean genotypes are photoperiod-sensitive and do not flower in long days. For example, in a field trial conducted in Southwestern Louisiana in 1982, 25 cultivars of P.tetragonolobus, one of P. scandens and two of P.palustris were sown in May. Flowering was observed to be strongly daylength-dependent with only 2 cultivars of P.tetragonolobus flowering in August; most cultivars did not flower until mid-to late September and some not until early November, thus limiting seed production (82). Genetic breeding for a successful search for day-neutral genotypes from germplasm collections is required (80).

Table IV

Winged bean varieties with greatest yield potential

Variety	Origin	Days to flowering	Days to maturity	Yield (kg/ha)			Recommendation
				seed	fresh tuber	dry pod	
Nakhon Sawan	Thailand	77	179	1623	1229	3193	A
LBNC1	Indonesia	87	190	1322	2854	2678	A
LBNC3	Indonesia	86	193	1282	4363	2151	A
UPS 121	Papua New Guinea	76	165	1349	881	2773	B
UPS 122	Papua New Guinea	68	162	1386	1202	3001	B

A - as seed, green pod and tuber crop.

B - as seed producer for all areas; may be used for green pod production.

Ref. (79).

A number of problems related to utilization of various portions of the plant for food and feed were recently reviewed by us
(61). Notably the hard seed coat, poor hydration capacity and
hard-to-cook phenomena needs to be overcome.
Preliminary studies on the potential of winged bean as a
multi-purpose legume cover crop in India (83) and Philippines (84)
have been reported. Data obtained on these field trials seem en-
couraging though they need further confirmation based on large
scale crop planting. One optimistic feature in winged bean research is that during the last few years, agronomical studies are
being conducted in countries such as Czechoslovakia (85), Belgium
(86,87), Italy (88), Brazil (89), Guatemala (90) and USA (91,92),
where winged bean has not been grown traditionally, even as a home
garden crop. Such a pooled effort is badly needed to fill the gaps
of knowledge which exist in the genetics, physiology and agronomy
of winged bean.
In conclusion, winged bean is a legume crop with great potential but a combined research effort by agronomists, plant breeders, physiologists and nutritionists is needed to eliminate the ob-
stacles which stand in the way of its realization. Large scale nutritional benefit from this legume awaits further research develop-
ments.

Acknowledgments

Financial assistance provided by the Nestle Foundation (Switzer-
land) as a scholarship grant to the senior author is gratefully appreciated.

Literature cited

1. Verdcourt, B.; Halliday, P.A. Kew Bull. 1978, 33, 191-227.
2. Allen, O.N.; Allen, E.K. "The Leguminosae: A Source Book of
   Characteristics, Uses and Nodulation"; University of Wisconsin
   Press, 1981, p. 556-58.
3. Smartt, J. Econ. Botany. 1984, 38, 24-35.
4. Blagrove, R.J.; Gillespie, J.M. Austr. J. Plant Physiol.
   1978, 5, 371-75.
5. Blagrove, R.J.; Gillespie, J.M. Trop. Grain Legume Bull.
   1978, 13, 40-44.
6. Blagrove, R.J.; Gillespie, J.M. Food Technol. in Austr. 1979,
   31, 149-50.
7. Gillespie, J.M.; Blagrove, R.J. Austr. J. Plant Physiol.
   1978. 5, 357-69.
8. Saio, K.; Nakano, Y.; Uemoto, S. Food Microstructure. 1983,
   2, 175-81.
9. Yanagi, S.O.; Yoshida, N.; Saio, K. Agric. Biol. Chem. 1983,
   47, 2267-71.
10. Yanagi, S.O. Agric. Biol. Chem. 1983, 47, 2273-80.
11. Okezie, B.O.; Martin, F.W. J. Food Sci. 1980, 45, 1045-51.
12. Garcia, V.V.; Palmer, J.K. J. Food Technol. 1980, 15, 469-76.
13. Hildebrand, D.F.; Chaven, C.; Hymowitz, T.; Bryan, H.H.;
    Duncan, A.A. Agronomy J. 1981, 73, 623-25.
14. Sathe, S.K.; Salunkhe, D.K. J. Food Sci. 1981, 46, 1389-93.

15. Sathe, S.K.; Deshpande, S.S.; Salunkhe, D.K. J. Food Sci. 1982, 47, 503-9.
16. Hettiarachchy, N.S.; Erdman, J.W. J. Food Sci. 1984, 49, 1132-35.
17. Sri Kantha, S. M.Sc.Thesis, University of Peradeniya, Sri Lanka, 1980; Field Crop Abstr. 1983, 36, 383.
18. Udayasekhara Rao, P.; Belavady, B. J. Plant Foods. 1979, 3, 169-74.
19. Narayana, K.; Narasinga Rao, M.S. J. Food Sci. 1982, 47, 1534-38.
20. Del Rosario, R.R.; Loxana, Y.; Noel, M.G.; Flores, D.M. Phil. Agric. 1981, 64, 143-53.
21. de Lumen, B.O.; Salamat, L.A. J. Agric. Food Chem. 1980, 28, 533-36.
22. de Lumen, B.O.; Belo, P.S. J. Agric. Food Chem. 1981, 29, 884-86.
23. Chan, J.; de Lumen, B.O. J. Agric. Food Chem. 1982, 30, 42-46.
24. Chan, J.; de Lumen, B.O. J. Agric. Food Chem. 1982, 30, 46-50.
25. Hettiarachchy, N.S.; Sri Kantha,S. Nutrisyon(Philippines), 1982, 7, 40-51.
26. Kortt, A.A. Trop. Grain Legume Bull. 1978, 13/14, 44-46.
27. Kortt, A.A. Biochem. Biophys. Acta. 1979, 577, 371-82.
28. Kortt, A.A. Biochem. Biophys. Acta. 1980, 624, 237-48.
29. Kortt, A.A. Biochem. Biophys. Acta. 1981, 657, 212-21.
30. Kortt, A.A. Qualitas Plantarum. 1983, 33, 29-40.
31. Tan, N.H.; Lowe, E.S.H.; Iskander, M. J. Sci. Food Agric. 1982, 33, 1327-30.
32. Tan, N.H.; Wong, K.C. J. Agric. Food Chem. 1982, 30, 1140-43.
33. Pueppke, S.G. Biochem. Biophys. Acta. 1979, 581, 63-70.
34. Sri Kantha, S.; Hettiarachchy, N.S. J. Natl. Sci. Council Sri Lanka, 1981, 9, 223-28.
35. Higuchi, M.; Suga, M.; Iwai, K. Agric. Biol. Chem. 1983, 47, 1879-86.
36. Higuchi, M.; Tsuchiya, I.; Iwai, K. Agric. Biol. Chem. 1984, 48, 695-701.
37. Higuchi, M.; Iwai, K. Agric. Biol. Chem. 1985, 49, 391-98.
38. Appukuthan, P.S.; Basu, D. Anal. Biochem. 1981, 113, 253-55.
39. Ikura, K.; Sasaki, R.; Chiba, H. J. Nutr. Sci. Vitaminol. 1983, 29, 161-67.
40. Raghunath, M.; Belavady, B. Nutr. Rep. Internat. 1979, 20, 701-7.
41. Cerny, K.; Kordylas, M.; Pospisil, F.; Svabensky, O.; Zajic, B. Brit. J. Nutr. 1971, 26, 293-99.
42. Kimura, T.; Satanachote, C.; Yoshida, A. J. Nutr. Sci. Vitaminol. 1982, 28, 27-33.
43. Gross, R. Qualitas Plantarum. 1983, 32, 117-24.
44. Wong, K.C. Proc.Malaysian Food Self Sufficiency 1975 Conference, 1976, pp. 103-16.
45. de Lumen, B.O.; Gerpacio, A.L.; Vohra, P. Poultry Sci. 1982, 61, 1099-106.

46. Cerny, K.; Addy, H.A. Brit. J. Nutr. 1973, 29, 105-7.
47. Cerny, K.; Hoa, D.Q.; Dinh, N.L.; Zelena, H. Proc.2nd Internat. Winged Bean Seminar, 1981, Columbo, Sri Lanka. (in press).
48. Blaise, D.S.; Okezie, B.O. Bakers Digest. 1980, 54, 22-24 & 33-35.
49. Okezie, B.O.; Dobo, S.B. Bakers Digest. 1980, 54, 35-41.
50. Nmorka, G.N.; Okezie, B.O. Cereal Chem. 1983, 60, 198-202.
51. Gandjar, I. Proc. 1st Internat. Winged Bean Seminar, 1978, Manila, Philippines, 1980, pp. 330-34.
52. Saio, K.; Suzuki, H.; Kobayashi, T.; Namikawa, M. Food Microstructure, 1984, 3, 65-71.
53. Omachi, M.; Ishak, E.; Homma, S.; Fujimaki, M. Nippon Shokuhin Kogyo Gakkaishi. 1983, 30, 216-20.
54. Jensen, M.W. Ph.D. Thesis, Purdue University, Indiana, 1980.
55. Sri Kantha, S.; Hettiarachchy, N.S.; Erdman, J.W. J. Food Sci. 1983, 48, 441-444.
56. Axelson, M.I.; Cassidy, C.M.; de Colon, M.M.; Hacklander, E.H.; Neruda, G.S.; Rawling, K.A. Ecol. Food and Nutr. 1981, 12, 127-37.
57. Hacklander, E.H. J. Home Econ. 1984, 76, 14-17.
58. Claydon, A. Qualitas Plantarum. 1983, 32, 167-77.
59. "The Winged Bean: A High Protein Crop for the Tropics", National Academy of Sciences, 1981, 2nd ed.
60. Khan, T.N. "Winged Bean Production in the Tropics", FAO, Rome, 1982.
61. Sri Kantha, S.; Erdman, J.W. J. Amer. Oil Chem. Soc. 1984, 61, 515-25.
62. Smith, O.B.; Ilori, J.O.; Onesirosan, P. Animal Feed Sci. and Technol. 1984, 11, 231-37.
63. D'Mello, J.P.F.; Acamovic, T.; Walker, A.G. Trop. Agric.(Trinidad), 1983, 60, 290-93.
64. Wyckoff, S.; Mak, T.K.; Vohra, P. Poultry Sci. 1983, 62, 359-64.
65. Jaffe, W.G.; Korte, R. Nutr. Rep. Internat. 1976, 14, 449-55.
66. Sri Kantha, S.; Erdman, J.W. unpublished data, 1985.
67. de Lumen, B.O.; Reyes, P.S. J. Food Sci. 1982, 47, 821-24.
68. Vaidehi, M.P.; Annapurna, M.L.; Gururaja Rao, M.R.; Uma, C.N. J. Food Sci. and Technol.(India). 1982, 19, 136-39.
69. Poulter, N.H. J. Sci. Food Agric. 1982, 33, 107-14.
70. Kortt, A.A.; Caldwell, J.B. J. Sci. Food Agric. 1984, 35, 304-13.
71. Hafez, Y.S.; Mohamed, A.I.; Herath, W. Nutr. Rep. Internat. 1984, 29, 253-61.
72. Hildebrand, D.F.; Chaven, C.; Hymowitz, T.; Bryan H.H. Trop. Agric. (Trinidad). 1982, 59, 59-61.
73. Masefield, G.B. Field Crop Abstr. 1973, 26, 157-60.
74. Ikram, A.; Broughton, W.J. Soil Biol. Biochem. 1980, 12, 77-82.
75. Ikram, A.; Broughton, W.J. Soil Biol. Biochem. 1980, 12, 83-87.
76. Ikram, A.; Broughton, W.J. Soil Biol. Biochem. 1980, 12, 203-9.
77. Cheeke, P.R.; Telek, L.; Carlsson, R.; Evans, J.J. Nutr. Rep. Internat. 1980, 22, 717-21.

78. Zomer, E.; Klein, D.; Rozhanski, M.; Er-El, A.; Joson, L.M.; Goldberg, I. Biotechnol. Letters. 1981, 3, 513-18.
79. Khan, T.N.; Speijers, J.; Edwards, C.S. "Adaptation of Winged Bean: Report on the First International Winged Bean Trials", International Council for Development of Underutilized Plants, Calif., 1984.
80. Smartt, J. Expl. Agric. 1985, 21, 1-18.
81. Anon. Winged Bean Flyer (Philippines). 1982, 4(2), 32-33.
82. Taylor, R.W.; Meche, G.A. 74th Ann. Progress Report Rice Expt. Station, Crowley, LA 1982, pp. 452-55.
83. Banerjee, A.; Bagchi, D.K.; Si, L.K. Expl. Agric. 1984, 20, 297-301.
84. San Juan, N.C.; Abad, R.G. Hort. Abstr. 1982, 52, 1227.
85. Buresova, M. Agric. Tropica et Subtropica, 1979, 12, 107-29.
86. Iruthayathas, E.E.; Vlassak, K.; Reynders, L. Z. Pflanzenernaehr Bodenkd. 1982, 145, 398-410.
87. Iruthayathas, E.E.; Gunasekeran, S.; Vlassak, K. Sci. Hortic. (Amsterdam), 1983, 20, 231-40.
88. Bistocchi, N.; Cagiotti, M.R.; Crosta, G. Ann. della Facolta di Agraria Univ. degli Studi di Perugia. 1979, 33, 389-404.
89. Lam-Sanchez, A.; Faggioni, J.L.M.; Kronka, S.D.N. Cientifica, 1981, 9, 137-44.
90. Jarquin, R.; Quezada, R.; Gomez-Brenes, R.; Bressani,R. Arch. Latin Amer. Nutricion, 1982, 32, 111-29.
91. Bryan, H.H.; Duncan, A.A. Proc. Amer. Soc. Hort. Sci. 1978, 22, 252-64.
92. Lynd, J.Q.; Lurlarp, C.; Fernando, B.L. J. Plant Nutr. 1983, 6, 641-56.

RECEIVED December 26, 1985

18

Tropical Seeds, Legumes, Fruits, and Leaves as Sources of Protein

Robert E. Berry

U.S. Citrus and Subtropical Products Laboratory, South Atlantic Area, Agricultural Research Service, U.S. Department of Agriculture, Winter Haven, FL 33883-1909

Many tropical plants are surprisingly high potential sources of protein with some leafy plants comprising as much as 40% dry-matter as protein, some fruits ranging as high as 40% protein and some aquatic weeds containing up to 5% protein. The amino acid compositions of many of these are well balanced, while others, particularly proteins from fruits, often need amino acid supplements or combinations with other protein sources. In extraction and use of leaf proteins, certain types of lipids provide interference and must be removed prior to or during the protein extraction process. In fruits and some tubers, caloric density becomes an important factor. Often, in order to obtain the recommended dietary allowance of protein from a specific fruit source, a higher than recommended amount of calories would need to be consumed. Thus, in order to be useful dietary sources of protein, some sources must be reduced in caloric content and/or supplemented with protein concentrates. While protein and amino acid composition of some water weeds are low compared with other leaves, they are easy and very low cost to produce. In many tropical areas they grow wild and would only need to be harvested. Natural production would be sufficient for supplying very large quantities.

Historically, a major form of malnutrition in most parts of the
world has been protein deficiency and this has been especially
prevalent in subtropical and tropical areas where 80% of the world
population will reside by the end of the next decade (1). While
legumes and leaves have been considered as potential protein sources
for many years, only recently have the protein potentials of less
obvious sources such as tropical fruits, some unique tropical seeds
and aquatic water weeds been considered. As the protein
deficiencies in world diets become more critical, increasing
attention will be given to new sources of protein which can be grown
in tropical and subtropical areas and which have not been given
significant attention in the past (2). While, in general, fruit and
vegetables provide only about 3-4% of the dietary protein in most
countries, this is more than doubled in a few such as Spain, Peru
and the Philippines (3). A review of the overall protein contents
of some tropical plants, the crude protein and extractable amounts
in comparison with average sizes and edible amounts of some fruits
and the potential production yields of some seeds and leaves,
indicates there is significant potential for additional protein as
dietary supplements in tropical areas. Past work on protein
extractability from leaves and grasses has indicated, while they are
good potential sources of leaf protein concentrates (LPC), some
lipid components interfere with protein extractability (4). The
concept of caloric density of foods is of importance when
considering certain proteins. The U.S. Senate Select Committee on
Nutrition and Human Needs recommended consumption of fruits and
vegetables should be increased, meats, fats and refined sugars
decreased, and about 12-14% of the total caloric intake should be
derived from protein (5). In many starchy crops (roots and tubers)
and fruits, the sugars and starches so dilute the protein content
that in order to be a significant protein source, too many calories
would have to be consumed. Thus, many foods of these kinds must be
consumed as complements to other dietary components higher in
protein content, or the protein from these foods must be extracted
and concentrated.

Many waterways of tropical areas are plagued with water weeds
which must continually be removed to keep lakes and streams
navigable and they generally cause a nuisance. These aquatic weeds
can also serve as potential protein sources. Though they have the
disadvantage of being relatively low in protein content, they have
the advantage of being widely available, easily grown and
practically no cost to produce.

Protein in tropical plants

The protein content of many tropical and subtropical plants is
surprisingly high. A study by Martin and Ruberte in 1975 (6)
included leaves of many plants which could be easily grown in
tropical areas and which contained about 20% or higher protein on a
dry-weight basis. In 1975, Hall et al. (7) determined amino acid
compositions of many of these plants by semi-quantitative thin-layer
chromatography. Their results, indicated in Table I, showed many
types of tropical plants have protein contents above 20%, with the
highest being the castor bean and the balsam pear.

Table I. Protein and Free Amino Acid Compositions of
Leaves from Tropical and Subtropical Plants
[Adapted from Hall et al. (7)]

Common Name	Type and Growth Habit[b]	Protein Content (Mean %)	Free Amino Acids[a]
Castor bean	P[c] herb	41.3	.165
Balsam pear	A[c] vine	32.5	.107
Cowpea	P vine	31.8	.182
Cassava	P shrub	31.6	.107
Purslane	A herb	27.5	.243
Chayote	P vine	24.4	.162
Mulberry	P tree	23.9	.128
Ceylon spinach	P vine	23.3	.152
Pigeon pea	P shrub	23.3	.081
Cup panax	P shrub	22.4	.091
Peperomia	A herb	21.6	.109
Banana	P herbaceous tree	20.9	.133
Dwarf Coral tree	P tree	20.4	.124
Coffee	P shrub	19.5	.076
Hibiscus	P shrub	17.5	N.D.[d]
Beach morning glory	P herb	16.3	N.D.
Poor man's orchid	P tree	15.5	N.D.
Indian mulberry	P tree	14.5	N.D.
Tamarind	P herb	13.0	N.D.
Bignay	P tree	11.4	N.D.
Baobab	P tree	10.8	N.D.
Mango	P tree	7.5	N.D.
Indian rubber	P tree	5.9	N.D.

[a]Milliequivalents of free amino acid per gm of dry leaves.
[b]P = Perennial.
[c]A = Annual.
[d]N.D. = Not determined.

Cow peas, cassava and pigeon peas also ranked high and these
plants are currently used as part of the staple diet in many
tropical countries. Their studies concluded that any leaf crop
with high protein content and significant essential amino acids
should be strongly considered as a potential source for leaf
protein concentrate (LPC). They also indicated such crops if not
already consumed, should be promoted for consumption as
supplementary foods. Their studies verified that leguminous
plants such as cowpea and pigeonpea could serve as good sources of
supplementary amino acids because of the high lysine content which
is commonly limiting in many diets based on cereals such as rice,
corn, and wheat (8). They also recommended that the nutritional
quality of cassava leaves (high in lysine and methionine) would
greatly enhance a diet of cassava root, a staple food in the
tropics. The castor bean has not been used as a protein source
primarily because of a poisonous alkaloid which combines with the
protein ricin in the mature leaves. They recommended studies to
solve this problem through improved processing techniques or plant
hybridization methods.

Protein in tropical fruits

While fruits are not normally considered potential sources of
protein, some have been found recently to be surprisingly high in
crude protein in the edible portion (9, 10, 11). As indicated in
Table II some of the more common fruit such as banana, persimmon
and mango have crude protein in the range of 1 g or less/100 g of
fruit flesh. However, some less widely known fruits such as Mamey
sapote, tucuma, and decne have crude protein contents above 2
g/100 g and baobab-seeds contained over 12 g/100 g.

Table II. Potential Protein in Some Tropical Fruits[a]

	Average Fruit Weight (g)	% Edible	%[b] Solids	Crude[c] Protein g/100 g of Fruit	Total Required Amino Acids mg/100 g Fresh Fruit
Baobab-seed[d]	114	35	90	12.56	–
-pulp[d]	56	68	90	3.31	–
Decne	714	90	30	4.58	–
Tucuma	20	19	45	2.67	1181
Mamey sapote	800	82	34	2.12	535
Avocado	85	57	21	1.61	666
Longan	6	52	19	1.31	301
Breadfruit[d]	2018	80	32	1.36	–
Banana[e]	257	68	24	1.00	–
Persimmon	20	60	24	0.62	234
Mango	550	69	15	0.42	428

[a]From Hall et al. (9).   [d]From Oliveria (10).
[b]100% - % moisture.   [e]From Adams (11).
[c]N x 6.25.

However, Oliveria (10) indicated the decne fruit contained a very
high moisture content and the baobab-seeds had a very high fiber
content. Thus, it would not be practical to expect one to consume
relatively large quantities of these fruits in the fresh form.
However, there are potential treatments by which the fiber content
could be lessened or the moisture content could be reduced. On a
per fruit basis, breadfruit, Mamey sapote and decne appeared to be
good protein sources because of the relatively high fruit weight,
as well as the significant crude protein content. According to
Hall et al. (9) the tucuma was one of the best apparent sources of
crude protein among the tropical fruits they studied. Because
these fruit are quite unusual, a brief description seems
appropriate. These are fruit from a type of palm grown around the
Amazon basin in South America and sold in that region. They are
usually harvested from wild stands and the fruit has a heavy,
starchy, semi-moist consistency and bland flavor. It has been
used in soups, stews, and cooked with other foods. While the
Mamey sapote and the avocado also contain significant crude
protein in the edible portions, they both contain very high
starch, fat, and carbohydrate and therefore are high in dietary
calories. Thus, they would be impractical to serve as primary
protein sources.

When quality of fruit proteins, (indicated by the content of
amino acids required in the human diet) is considered several are
significant, as indicated in Table III. This lists, from left to
right in descending order, the amounts of required amino acids in
several fruits, recently reported. Among the fruits reported
here, the tucuma and avocado provide a good balance and a high
quantity of the required amino acids. Almost all fruits reported
here were relatively low in the sulfur containing amino acids,
particularly cysteine and cystine. However, dried date, tucuma,
and Mamey sapota were relatively high in amounts of these amino
acids as well. Most of the more common fruits such as peach,
orange and mango were relatively low in sulfur containing amino
acids, as well as other required amino acids. As indicated by
these studies many fruits would make excellent dietary supplements
to improve content of certain amino acids.

Protein source vs production potential

Where arable land is high cost, high value, or in scarce supply
the production potential of protein becomes an especially
important factor. Fruits and leaves from wild, undomesticated
plants provide attractive sources of protein in the diet because
of their natural acceptability by local inhabitants and their
wide-spread accessibility due to native growth. However, for
domesticated and commercialized situations the production of
protein per unit area of arable land is an important factor.
Seeds, leaf crops, and grasses are very efficient protein
producers from the standpoint of amount per unit area. As
indicated in Table IV, the more commonly used seeds and legumes,
soybeans and wheat have relatively high protein production
potentials (around 400 to 800 kg/ hectare).

Table III. Dietary Required Amino Acids (mg/100 g Fresh Fruit) in Some Fruits
[Adapted from Hall et al. (9)]

Amino acid	Tucuma	Avocado	Dried Date[a]	Mamey Sapota	Longan	Peach	Orange	Mango
Isoleucine	125	72	66	46	26	13	23	20
Leucine	201	126	114	84	54	29	22	32
Lysine	135	106	81	84	46	30	43	28
Methionine	33	28	22	18	13	31	12	7
Cysteine + Cystine	28	4	52	18	3	9	10	4
Phenylalanine	107	74	74	53	30	18	30	19
Tyrosine	145	50	21	55	25	21	17	11
Threonine	130	75	76	58	34	27	12	20
Valine	153	109	93	77	58	40	31	29
Histidine	125	21	33	42	12	17	12	13
Total	1181	666	632	535	301	235	212	182

[a]From FAO (20).

Table IV. Production Potential of Protein From Some
Seeds and Leafy Vegetables [from Berry (2)]

	kg Protein /ha
Seeds	
Soybean (oilseed)	510[a], 785[b]
Wheat (cereal seed)	392[b]
Cowpea (legume seed)	330[c]
Urd bean (legume seed)	254[d]
Leaves	
Alfalfa (Medicago sativa)	3000[e], 2690[b]
Sorghum x sudan grass hybrid	2320[b]
(Sorghum sudanesis)	
Transvala digitgrass	2179[f]
(Digitarea decumbenc)	
Tetrakalai	1760[g]
(Phasedus aureus)	

[a]From Food Ind. S. Aft. (1).   [e]From Singh (22).
[b]From Stahmann (19).   [f]From Sotomayor-Rios et al. (23).
[c]From FAO (20).   [g]From Kohler et al. (24).
[d]From Jeswani (21).

Cow pea and urd bean, two common legumes are only slightly lower.
An excellent review by H. D. Tindall of the leguminosae plants
indicates most of their seeds have high protein contents, are
directly consumable, and they are easily grown in tropical and
subtropical climates (30).
    However, neither seeds nor fruit can approach the extremely
efficient production of leafy plants and most forage grasses.
Alfalfa, one of the most commonly grown sources, produces around
3,000 kg/hectare. Certain grasses such as the sorghum x sudan
grass hybrid and the transvala digitgrass indicated in Table IV
have relatively high protein but not as high as alfalfa. These
grasses are some of the better prospects among those surveyed at
the USDA Tropical Horticultural Laboratory in Rio Piadres, Puerto
Rico where many grasses, leaf crops and hybrids are being
studied. It is notable that leaves produced more crude protein
than seeds by factors of 6 to 8 times and leafy vegetables yield
the highest protein/unit area. As reported by Abbott in 1966 (3),
one acre of land can produce in three to four months about eight
tons of fresh leafy vegetables (grasses especially) containing up
to 350 pounds (160 kg) of protein. Legumes have a much lower
yield and in fact, peas and beans rank below some starchy crops
like potatoes and corn in quantities of protein produced per unit
of land area (12).
    Another point to be considered is the form in which the
protein-containing food is consumed and the efficiency of
extraction or use of the protein in human diet. When this is
considered, most of the seed-type foods, beans and legumes, have
efficiencies which overcome their lowered production potential.
Thus, these foods can be directly consumed and this results in a
very efficient use in the human body. Many grasses and

leafy-type vegetables must be either extracted and concentrated or
fed to animals and protein consumed by eating the animal. Protein
can usually be more easily extracted from beans and seeds than
from leaves. However, in an extensive study, Martin and Ruberte
(6) indicated there are many tropical leaves that can be consumed
directly and serve as excellent sources of efficient protein in
the human diet.

Protein in leafy vegetables

As pointed out by Martin and Ruberte (6) of many vegetables
normally consumed such as beans, seeds, peas, and tubers, the
leaves contain significantly more protein than the part commonly
consumed. They reported that the % crude protein in many leaves
ranges as high as 20% in bamboo, curry, and lettuce to greater
than 30% in leaves from cassava, cowpea, balsam pear and pumpkin.
Cassava, taro, okra, soybean and pigeon pea are good examples of
protein content of leaves being higher than that of seed, pod, or
tuber.
  Many studies have been reported on extractability and
concentration of protein from leafy plants (13, 14). While it is
more efficient to consume the leafy vegetable or fruit directly,
considerable quantities of less available protein can be made
available through extraction and concentration into LPC. As
pointed out by Pirie (13) and Kohler & Knuckles (14), an LPC
processing plant could be oriented near packinghouses for leafy
vegetables so that leaves, stems, and stalks trimmed from these
vegetables before being sent to market, could subsequently be
extracted for protein. Leaf protein processes are covered in more
detail in another chapter.
  The protein in four different tropical leaves and the yield of
crude and final LPC were studied by Nagy et al. in 1978 (15).
They found that the crude protein in cassava leaves and sorghum x
sudan hybrid grass were relatively high (Table V) but the dried
yield of true leaf concentrate was much higher from sauropus.

Table V. Weights and Protein Contents of Leaves from Four
          Tropical Plants [from Nagy et al. (4)]

Tropical Leaves	Fresh Wt Leaves g	Dry Content g	Crude Pro.[a] Dry Leaves, g	Crude[b] LPC Prot., g	True[b] LPC Prot., g
Chaya	873	170	41.3	15.1	7.2
Sorghum x sudan	1230	180	18.7	11.1	4.7
Cassava	561	102	18.0	10.2	5.7
Sauropus	1216	281[c]	83.7	21.4	10.5

[a] N x 6.25.
[b] Dried weight.
[c] Contains woody stems.

A problem encountered especially in the LPC from grasses and some types of leaves is the interference of lipids extracted with the protein concentrate which were difficult to separate. As indicated in Table VI, although the protein from grasses is considered highest quality from plants, based on required amino acids, that from other tropical leaves such as cassava, Chinese cabbage and Tamu-TEX-sci (an experimental grass hybrid from the University of Texas) all compare favorably. These plants all show a good balance of amino acids required for humans and especially some of those such as methionine which are usually low in plant protein. However, they all have a relatively low level of threonine and the experimental grass hybrid is low in phenylalanine. Three of these plants, Chinese cabbage, the grass hybrid and cassava leaves can be consumed directly as human food, as well as used as a source of LPC.

Regarding the interference of lipids and fatty acids in preparation of LPC, Nagy et al. (15) made an extensive study of this problem and determined, as indicated in Table VII, that the lipid content and total lipid distribution in some green protein fractions is indeed significant and can present a problem with protein extractability and purification. They indicated however, most of the lipid appeared to come from extraction of cell walls and ruptured cellular contents during the maceration process. They also suggested the lipids were released and became extractable primarily through enzymatic reactions after maceration of the leaves for extraction. Thus, they postulated that the time interval between maceration of the plant tissue and the application of heat to coagulate proteins was critical. They recommended heat inactivation of enzymes and precipitation of proteins as quickly as possible after maceration to minimize this interference.

Role of caloric density in protein effectiveness

A serious problem with some vegetables and particularly with fruit as a source of protein is the dilution of protein with carbohydrates so that an inordinate amount of calories must be consumed to obtain the needed amount of protein and amino acids for human growth. A survey of calories and protein content/100 g fresh weight of several different leaves, pods, seeds and tubers is shown in Table VIII. This indicates the desirability of certain leaves, pods and seeds over tubers, for example, as a protein source.

Foods that yield in the range of 5-10 calories/g protein should be excellent dietary sources of protein. However, those that yield 50-100 or more calories/g protein should probably be used more where calories for energy are needed and supplemented with other protein sources. In many countries man has supplemented his diet with protein by the consumption of animal products and meat in order to eliminate this problem and balance the protein-caloric values.

Table VI. Required Amino Acids in Some Tropical Leafy Vegetables and Grasses [g Amino Acids/100 g Protein; Adapted from Nagy et al., (4)]

	Brassica Chinensis[a] (Chinese Cabbage)	Brassica Carinata[b] (TAMU-Tex Sci)	Cassava[c]	Alfalfa[d]	Barley[a]
Lysine	7.1	6.0	6.7	6.3	6.6
Phenylalanine	6.2	2.5	5.9	5.8	6.2
Methionine	1.9	1.4	1.6	2.0	2.2
Threonine	5.2	5.0	4.8	5.0	5.1
Leucine	9.2	9.5	9.9	8.9	9.3
Isoleucine	5.2	5.0	5.2	5.2	5.0
Valine	6.4	6.3	6.6	6.4	6.4
Tryptophan	1.6			1.7	

[a] From Byers (25).
[b] From Brown et al. (26).
[c] From Otoul (27).
[d] From Bickoff et al. (28).

Table VII. Lipid Distributions of Green Protein Fractions[a]

Tropical Leaves	Total Extracted Lipids g	Lipid/ Crude Protein	Total Lipid Distribution, %		
			Nonsaponifiables	Fatty Acids	Residue
Chaya	2.8	0.19	27.6	33.8	38.6
Sorghum x sudan	2.5	0.23	38.5	35.3	36.2
Cassava	1.2	0.12	25.2	32.0	42.8
Sauropus	4.8	0.22	30.4	34.1	35.5

[a] Adapted from Nagy et al. (4).

Table VIII.   Protein and Caloric Value of Different Types of
Vegetables (Per 100 g Fresh wt)[a]

	Calories	Protein (g)	Cal/ g. Prot.
Leaves			
Cassava	53	7.0	7.6
Tropical spinach	44	4.0	11.0
Kankong	17	3.0	5.6
Green vegetables (avg)	22	2.4	9.2
Yellow vegetables (avg)	18	1.5	12.0
Pods and seeds			
Beans (pods)	33	2.2	15.0
Asparagus bean (pods)	18	2.0	9.0
Peas (seeds)	48	3.4	14.1
Tubers			
Cassava	131	0.7	187.0
Sweet potato	121	2.0	61.0
Cocoyam	88	2.0	44.0

[a] Adapted from Terra (29).

Aquatic weeds as sources of plant protein

While many types of plant sources are attractive as sources of
protein, perhaps the most attractive of all would be those plants
which are a nuisance, grow wild rapidly and whose removal would be
an advantage to mankind. Roe and Bruemmer (16) studied the
extractable protein as percent of dry-matter from several aquatic
weeds which had become a nuisance in Florida and other tropical
and subtropical areas of the U.S. These and similar varieties are
also wide-spread among most tropical areas of the world. As
indicated in Table IX they found protein contents ranging from .3
to 5.4 % of the dry-matter. A further study indicated a higher
yield of protein could be obtained when dimethyl sulfoxide (DMSO)
was used for extraction. Their studies also indicated, (Table X)
that the protein from these water-weeds had a relatively good
balance of amino acids, being low mainly in methionine. It was
also observed that those extracts made with DMSO, although
yielding generally higher amounts of protein and individual amino
acids, yielded much lower amounts of the sulfur-containing
cysteine, cystine and methionine. They speculated that the DMSO
probably decomposed the sulfur-containing amino acids due to its
chemical affinity for sulfur groups. They also noted a lower
amount of lysine in those extracts from DMSO and, although they
could not explain this completely, it may also be due to reaction
between the basic amino acid and DMSO.

Table IX. Extractability of Protein from Aquatic Weeds[a]

	Dry Matter (DM) %	Extractable Protein % of DM
Illinois pondweed	13.3	5.4
Water hyacinth	10.6	1.8
Water lettuce	6.9	0.9
Hydrilla	7.8	0.6
Cattails	27.5	0.3

[a]Adapted from Roe and Bruemmer (16).

Table X. Amino Acid Compositions of Acid Hydrolyzed
Protein Fractions from DMSO and Aqueous Extracts of
Pondweed and Water Spinach[a]

	Amino Acid Content: g/100 g Recovered			
	Pondweed		Water Spinach	
Amino Acid	DMSO	Aqueous	DMSO	Aqueous
Lysine	3.40	7.03	2.89	6.63
Histidine	3.17	2.59	5.06	2.43
Arginine	2.43	5.33	2.67	5.88
Aspartic	10.27	10.29	10.08	10.12
Threonine	5.99	5.33	6.44	5.38
Serine	6.22	5.50	5.42	4.99
Glutamic	11.77	10.77	12.33	11.11
Proline	5.09	4.61	5.12	4.70
Glycine	7.35	5.24	6.44	5.45
Alanine	7.33	5.87	7.22	5.77
Cystine (1/2)	0.30	3.54	0.29	3.64
Valine	7.21	6.32	7.36	6.25
Methionine	0.22	2.10	00	2.05
Isoleucine	6.16	5.18	6.34	5.00
Leucine	11.25	8.71	10.97	8.56
Tyrosine	2.63	4.60	2.53	4.86
Phenylalanine	7.28	5.40	7.07	5.78

[a]Adapted from Roe and Bruemmer (16)
and FAO (18).

Bruemmer and Roe's finding (17) that hot DMSO could extract 8%
of the dry-matter of water spinach as protein should encourage
agriculturalists and sanitation engineers interested in the
economic aspects of aquatic plants. These are often grown to
control the nutrient run-off in lakes from agricultural lands and
waste-treatment plants. Water spinach grows prolifically and acts
as a living sponge for removing the water soluble nutrients from
lakes and streams. Cattails, water hyacinths, water spinach and
hydrilla drastically lower the nitrite-nitrogen content of water
streams passed through their beds, or in which beds of these
plants are grown. They are currently used in field tests in

several waste disposal systems as methods to remove nutrients from
drainage water and waste treatment streams. The discovery of ways
to increase the extractable plant protein with a potential food or
animal feed use for these plants should greatly enhance the
economic feasibility of using them to control waste and excess
nutrients in water streams. The capability of such plants to
concentrate potentially toxic metals would have to be considered
and carefully controlled.

Conclusions

Seeds, legumes and leaves are among the best sources of potential
low-cost protein, considering the efficiency with which they can
be used by the human body upon direct consumption, coupled with
their protein production potential per unit area of arable land.
While proteins from these sources are very similar in amino acid
composition, a few cultivars showed notable variations in
methionine content and a well-balanced diet should contain
portions of several of those cultivars, perhaps mixed with certain
fruits as well. Some lesser known tropical fruits contain
significant amounts of proteins and almost all fruits contain
fairly well-balanced proteins considered from the content of
essential amino acids. However, in considering fruits and tubers
as sources of protein the caloric density of the food must be
carefully considered so that protein/calories is not too low due
to high calorie carbohydrates and/or lipids. Aquatic weeds and
certain nuisance plants, while not widely considered as protein
sources, nevertheless have significant amounts of protein which
with the use of specialized extraction techniques such as
extractability with DMSO, yield new sources of protein. Although
contents of these plants are not as high as those grown
specifically for foods, they offer the additional attractions of
being easily grown, low cost to grow, widespread in availability
and a great advantage for clean-up of waste-waters and streams by
harvesting and using them as a practical source for leaf protein
concentrate.

Literature Cited

1. U.N. Items 8 and 9 of the Provisional Agenda E/Conference,
   United Nations World Food Conference, Rome, 1974.
2. Berry, R. E. Food Technol. 1981, 35(11), 45-49.
3. Abbott, J. C. In "World Protein Resources"; ADVANCES IN
   CHEMISTRY SERIES No. 57, American Chemical Society:
   Washington, D.C., 1966; p. 1-14.
4. Nagy, S.; Telek, L.; Hall, N.T.; Berry, R. E. J. Agric. Food
   Chem. 1978, 26(5), 1016.
5. Peterkin, B. B. Food Technol. 1978, 32(2), 34.
6. Martin, F. W. and Ruberte, R. M. In "Edible Leaves of the
   Tropics"; Antillian College Press: Mayaguez, Puerto Rico,
   1975.
7. Hall, N. T.; Nagy, S.; Berry, R. E. Proc. Fla. State Hortic.
   Soc.,1975, 88, p. 486.
8. Betscart, A. A.; Kinsella, J. E. J. Agric. Food Chem. 1974,
   22, 116-122.

9. Hall, N. T.; Smoot, J. M.; Knight, R. J. Jr.; Nagy, S. J. Agric. Food Chem. 1980, 28, 1217.
10. Oliveria, J. F. Ecology of Food and Nutrition 1974, 3, 237.
11. Adams, C. F. "Nutritive Value of American Foods," Agricultural Handbook #456, U. S. Dept. of Agric., 1975.
12. Milner, M. J. Agric. Food Chem. 1974, 22(4), 458.
13. Pirie, N. W. Nutr. Clin. Nutr. 1975, 1, 341.
14. Kohler, G. O.; Knuckles, B. E. "Edible proteins from leaves." Food Technology 1977, 31(5) 191.
15. Nagy, S.; Nordby, H. E.; Telek, L. J. Agric. Food Chem. 1978, 26(3) 701-706.
16. Roe, B.; Bruemmer, J. H. Proc. Fla. State Hort. Soc., 1980, 93, p. 338-340.
17. Bruemmer, J. H.; Roe, B. Proc. Fla. State Hort. Soc., 1979, 92, p. 140-143.
18. FAO, "Amino Acid Content of Foods and Biol. Data on Protein, "U.N. Food and Agriculture Organization, Rome, 1970.
19. Stahmann, M. A. Econ. Bot. 1968, 22, 73.
20. FAO, FAO Agricultural Studies, No. 55, U. N. Food and Agriculture Organization, Rome, 1961.
21. Jeswani, L. M. "Food Protein Sources"; Pirie, N. W., Ed.; Cambridge Univ. Press: Cambridge, 1975; p. 13.
22. Singh, N., Central Food Technological Research Institute Report, Mysore, India, 1967.
23. Sotomayor-Rios, A.; Velez-Santiago, J.; Torres-Riveria, S.; Silva, S. J. Agric. Univ. P.R.,1976, 60, 294.
24. Kohler, G. O.; Bickoff, E. M.; deFremery, D. Univ. Calif. Spec. Publ. 3058, 1976, p. 116.
25. Byers, M. "Leaf Protein: Its Agronomy, Preparation, Quality and Use"; Pirier, N. W., Ed.; Blackwell Scientific Publications: Oxford, 1971.
26. Brown, H. E.; Stein, E. R.; Saldana, G. J. Agric. Food Chem. 1975, 23, 545.
27. Otoul, E. Bull. Rech. Agron. Gembloux. 1974, 9(2), 159.
28. Bickoff, E. M.; Booth, A. N.; de Fremery, D.; Edwards, R. H.; Knuckles, B. E.; Miller, R. E.; Saunders, R. M.; and Kohler, G. O. "Protein Nutritional Quality of Foods and Feeds"; Friedman, M., Ed.; Marcel Dekker; New York, 1975, p. 319.
29. Terra, G. J. A. Tropical Vegetables Comm. 54e, Ept. Agric. Res. Royal Tropical Inst., Amsterdam, Holland, 1966.
30. Tindall, H. D. In "Vegetables in the Tropics", The leguminosae; AVI Publishing Co.: Westport, Connecticut, 1983; pp. 250-321.

RECEIVED December 26, 1985

19

Protein of the Sweet Potato

W. M. Walter, Jr.[1], and A. E. Purcell[2]

[1]Agricultural Research Service, U.S. Department of Agriculture, and North Carolina Agricultural Research Service, Department of Food Science, North Carolina State University, Raleigh, NC 27695
[2]Department of Food Science and Nutrition, Brigham Young University, Provo, UT 84602

The sweet potato ranks sixth in average yearly
production among the world's major food crops.
The crude protein content ranges from 1.3% to >
10% (dry weight basis). Significant potential exists
for increasing the protein content by
breeding/selection and optimization of production
practices. From 60-85% of the nitrogenous material
is protein, and the remainder is mostly amino and
amide nitrogen. Humans have been maintained in
nitrogen balance using sweet potato as the major
source of nitrogen. The protein efficiency ratio
(PER) for isolated sweet potato protein is equal
to that of casein. Heat processing lowers lysine
bioavailability, dependent upon the severity of
the heat treatment and the amount of reducing sugar
present during heating.

The sweet potato (Ipomoea batatas L.) is an important contributor
to human nutrition in many parts of the world. Sweet potato ranks
sixth in annual world production at 137 million metric tons
(1975-1977) (1) behind wheat, rice, maize, potato, and barley.
Although starchy roots are generally considered to provide only
calories to the diet, the sweet potato provides 73% of the
required protein per calorie (2, 3) for an adult male. The average
yield for sweet potatoes for 1975-1977 (1) was 9,621 kg/ha, making
it second only to white potatoes among the ten leading crops
produced worldwide. There is significant potential for increased
yields, provided production practices are optimized and high
yielding cultivars are grown. In the United States, for example,
the mean yield in 1980 was 13,108 kg/ha (4). High yields and
a 110-130 day growing season make the sweet potato an attractive
source of calories and other nutrients for tropical regions of
the world. It is noteworthy that the majority of the countries
with an annual income of less than $500 (US) per capita are
located in the tropics. Thus, the sweet potato is potentially

0097-6156/86/0312-0234$06.00/0
© 1986 American Chemical Society

an outstanding candidate for increased production in this area. Although not an important source of protein in the United States, the sweet potato is consumed extensively in New Guinea, and in parts of that country, provides up to 40% of the crude protein in the diet (5).

Data are not available for protein production worldwide. However, an estimate of the protein contribution provided by sweet potatoes can be made if we assume a mean dry matter content of 28% and a mean protein content of 5%. Based on these assumptions, the sweet potato provides 1.92 million metric tons of protein worldwide. The yield of protein would be 134 kg/ha using worldwide yield values or 184 kg/ha using US production values.

Sweet Potato Protein

The diet must provide those amino acids which the body cannot synthesize (essential amino acids, EAA) and nitrogen in the form of nonessential amino acids (NEA). Both EAA and NEA are required for biosynthesis of proteins and other nitrogen-containing compounds necessary for homeostasis or growth. Thus, the total nitrogen content of a specific food must be considered to be nutritionally significant.

For those sweet potato cultivars studied, the crude protein (N x 6.25) contains both protein and nonprotein nitrogen (NPN). The NPN content has been demonstrated to range from 15 to 37% at harvest (6, 7). The only published report of the composition showed the NPN fraction to be nutritionally unbalanced, containing mostly amino acids and amides (6). The major components were asparagine, 61%; aspartic acid, 11%; glutamic acid, 4%; serine, 4%; and threonine, 3%. Eighty-eight percent of the NPN fraction was accounted for by amino acids and amides. During the early part of storage, the NPN fraction decreased, then increased (8). The nonlinear nature of the change in NPN, coupled with the fact that nitrogen content decreased during storage, indicated that this fraction is part of a metabolically active nitrogen pool (9) and that the appreciable amount of nitrogen stored as asparagine is available for metabolic demands of the root. Although the NPN fraction of sweet potato is available to satisfy nitrogen requirements, only small amounts of EAA are present in this fraction.

The initial report on the nature of sweet potato protein indicated that most of the protein was a globulin "ipomoein" (10). The authors also stated that upon storage of the root, ipomoein was partially converted into a polypeptide which was considerably different from the parent material both in its chemical composition and its physical properties. Later workers using modern techniques reported the major soluble protein was a 25 k Da molecule (11). Only small amounts of this protein were found in roots stored for 1 year, suggesting that this protein is readily metabolized and is probably the storage protein. In addition, a second major protein identified as beta amylase was also shown to be minimally present in roots stored for 1 year.

Sweet potato protein is unequally distributed within the

root. The crude protein content is slightly greater in the stem
end than the root end. The only region which has been shown to
contain much higher protein levels is the outer layer adjacent
to the skin corresponding to precambial tissue (12, 13). Scraping
the roots removed ≥2.5% of the fresh weight (FW) and decreased
the root protein content by 4.4%, while a more drastic peeling
which removed ≥8.5% of the FW lowered the protein content by
12% (13). The tissue removed with the scrapings constituted 2.5%
of the total weight and contained 87% more protein per unit weight
than did the remaining tissue. The tissue removed by the deep
peeling treatment contained 47% more protein per unit weight
than did the tissue remaining after peeling. The above data
indicate that although the surface layers of tissue are
significantly higher in protein content than the underlying
tissue, the absolute amount of protein-rich material is small.
Consequently, it is not feasible to increase the protein content
by selective removal of tissue.

A protein concentrate can be obtained from sweet potato
roots (14). The laboratory method involved grinding with three
parts of water, screening to remove coarse fibrous material,
settling the starch, coagulating the chromoplasts, and
precipitating the protein. Sweet potatoes have been used as a
commercial source of starch and are still being used as such
in Japan (15). Commercial production of starch involves the first
three steps, i.e., grinding, screening and settling the starch.
It would appear that commercial quantities of sweet potato protein
might be readily available as a by-product of the starch industry.
The laboratory concentrates were bland, light-colored powders
containing 80-88% protein.

Crude Protein Variability

The sweet potato is a perennial, propagated vegetatively as an
annual for agricultural purposes. The plant is heterozygous and
is a hexaploid with a somatic chromosome number of 90. As would
be expected, genetic potential for variation in protein content
is great. Various workers have reported a protein range of from
1.3% to >10% (dry weight basis) (16, 17, 18), depending upon
the cultivar. There appears to be potential for increasing the
protein content by breeding, since the sweet potato has responded
quite well to selection for other traits when genetic variability
is present. Increase in protein content by selection is especially
important because many parts of the tropics, which are in need
of additional protein sources, consistently produce sweet potatoes
with low (< 4%) protein content (dry basis). Li (19) demonstrated
that a mass selection technique was effective in increasing crude
protein content and maintaining a high yield. A later study (7)
showed that NPN percent and trypsin inhibitor activity did not
increase as the sweet potato protein content increased. There
appeared to be some deterioration in the protein nutritional
quality with an apparent decline in relative amounts of valine,
aromatic amino acids and sulfur-containing amino acids. It should
be noted, however, that sample to sample variability among amino
acids is very great, and thus, more research is needed in this
area before a definite relationship can be determined.

Within cultivar variation of sweet potato crude protein
is high. Purcell et al. (20) reported a 13% coefficient of
variability between roots from a single hill and a 13% coefficient
of variability between hills in a single field. Field to field
variability was very great with Jewel cultivar, ranging from
3.99 to 8.81% protein (dry basis), depending upon the field
location. In a carefully controlled study, Collins and Walter
(21) reported that for six sweet potato genotypes grown at six
locations for 3 years (18 environments), protein content varied
in a statistically significant manner ($P < 0.01$) by genotype,
environment and the environment-genotype interaction.

Another study (22) of genotype-environment interaction for
sweet potatoes grown in the southern highlands province of Papua,
New Guinea, reinforced the finding of Collins and Walter (21)
with regard to the variability in crude protein content. The
data of Bradbury et al. (22) for 10 cultivars from 5 environments
showed a mean crude protein content of 1.51% (fresh weight) with
a standard deviation of 0.54%, a coefficient of variability of
35.8% (Table I). The gradient referred to in Table I was obtained
by plotting the mean crude protein content for the 5 environments
(bottom row, Table I) against the crude protein content for each
cultivar in each environment. The gradient or slope of the
resulting line provided a measure of the response of a given
cultivar to varying environments. The greater the gradient or
slope, the more the cultivar is affected by environment. From
Table I, it is apparent that the cultivar 'Simbul Sowar' is least
responsive to environment and still is high in crude protein
content. On the other hand, cultivar 'Takion' has the highest
mean crude protein content but much more environmental
instability. This type analysis is a valuable tool for improvement
of the crude protein content through cultivar selection.

Cultural practices also can affect sweet potato protein
content. Purcell et al. (23) reported that increasing amounts
of nitrogen fertilization up to 112 kg/ha caused an increase
in protein content but no change in the NPN. Neither sulfur nor
potassium influenced the protein content. Similarly, Constantin
et al. (24) found that nitrogen fertilization up to 67.3 kg/ha
linearly increased crude protein content. Kimber (25) reported
that when available nitrogen no longer affects yields, protein
content of the roots continues to increase. Other workers have
demonstrated that crude sweet potato protein content can be
increased through cultural management practices (26, 27). Length
of the growing season also has an effect on crude protein content.
Purcell et al. (28) found that the protein content decreased
0.0067% per day between 102 and 165 days. Concomitantly, dry
matter decreased linearly at 0.233% per day. In addition to
nitrogen fertilization rate and length of growing season, high
rates of irrigation caused decreases in both dry matter and
protein content (29). The results reported by Dickey et al. (7)
and Bradbury et al. (22) reinforce the concept that protein
content is not a reciprocal function of dry matter content. It
appears then that natural genotypic variability in crude protein
content provides a promising avenue to improve protein levels

Table I. Crude Protein Content (% Fresh Sweet Potato) of Ten Cultivars From Upper Mendi Grown in Different Environments

Name of Cultivar	Kiburu, 1982 Season[a]	Erave, 1982 Season[b]	Growth Conditions Upper Mendi, 1981 Season[c]	Upper Mendi, 1983 Season	Upper Mendi, 1983 Season Gypsum Added[d]	Mean	SD	Gradient (See Text)
Hopomehene (HO)	1.87, 2.06, 1.97	2.31, 2.37, 2.34	0.50	1.19	1.06	1.41	0.74	1.61
Takion (TA)	2.37, 3.00, 2.69	1.87, 2.25, 2.06	2.06	1.00	1.25	1.81	0.68	1.14
Soii (SO)	1.12, 1.44, 1.28	2.56, 1.37, 1.97	1.38	0.88	1.13	1.33	0.41	0.93
Sapel (SA)	1.81, 1.94, 1.88	1.69, 1.06, 1.38	0.75	1.06	1.69	1.35	0.46	0.31
Kariap (KA)	1.62, 1.37, 1.50	1.94, 1.87, 1.91	2.00	0.81	0.88	1.42	0.56	0.94
Pulupuri (PU)	1.12, 1.56, 1.34	2.12, 1.81, 1.97	1.00	1.31	0.81	1.29	0.45	1.00
Kariko (KO)	0.94, 0.87, 0.91	2.19, 2.44, 2.32	1.63	0.94	1.06	1.37	0.60	1.10
Simbul Sowar (SI)	1.69, 1.50, 1.60	1.87, 2.00, 1.94	1.69	1.31	1.50	1.61	0.23	0.51
Wanmun (WA)	2.37, 2.06, 2.22	2.31	0.88	1.94	1.19	1.71	0.64	1.12
Tomun (TO)	1.44, 1.69, 1.31, 1.48	2.81	1.56	1.56	1.31	1.76	0.60	1.33
Mean	1.60	2.10	1.35	1.20	1.19	1.51	0.45	

[a] Two (or three) roots from same plant and mean. [b] Roots from two different plants and mean. [c] From Bradbury et al. (13). [d] Gypsum added to soil at a rate of 500 kg/ha. From Bradbury et al. (22).

via selection. Selection for high protein cultivars which are
relatively insensitive to environmental differences and
optimization of cultural practices are also attractive research
areas for increasing protein content.

Nutritional Value

Feeding Studies. Although sweet potatoes are a significant source
of calories in many parts of the world, very little information
is available concerning the nutritional quality of sweet potato
protein as determined by controlled feeding studies in humans.
This is in striking contrast to numerous reported studies on
the feeding of white potatoes to humans (30). An early study
in which the sweet potato was used as the sole source of nitrogen
in the diet of humans was that of Adolph and Liu (31). They
reported that nitrogen balance could be maintained with sweet
potato nitrogen provided sufficient amounts were consumed.
Research by other workers (32, 33) also suggested the sweet potato
protein is readily utilized by humans.

Large amounts of sweet potato must be eaten to provide enough
nitrogen. Oomen (34) reported that in New Guinea, where 80-90%
of the total calories were obtained from sweet potato, the
subjects studied were usually in significant negative nitrogen
balance. Since negative nitrogen status means continuous breakdown
of body protein leading to serious malnutrition, Oomen (34) was
puzzled because the subjects seemed to be in good health. As
a result, he suggested that eating large amounts of sweet potato
might induce an intestinal microflora which was able to fix
gaseous nitrogen so that it could be utilized to synthesize amino
acids. Obviously, if such were the case, much of the knowledge
of protein nutrition would be in doubt since the validity of
nitrogen balance studies upon which most of this knowledge is
based would be in doubt. A later study (35) using carefully
controlled conditions indicated that both adolescent and young
adult males maintained in slightly negative nitrogen balance
through use of sweet potato as the major nitrogen source developed
clinical symptoms of mild protein malnutrition. These included
abnormal plasma free amino acid patterns and a decrease in
physical stamina. In addition, no evidence of in vivo nitrogen
fixation could be detected in fecal material, indicating that
the microflora induced by long-term consumption of sweet potatoes
are not capable of fixing nitrogen. The report that habitual
sweet potato eaters are somewhat independent of dietary nitrogen
appears to have no basis in fact.

Results reported by Huang et al. (35) indicated that with
teenagers a positive nitrogen balance could be maintained with
an intake of 0.67 to 0.71 g protein/kg body weight, where the
sweet potato furnished most of the protein. The energy requirement
for this level of protein consumption was 54 k cal/kg body weight.
The apparent protein digestibility was found to be 66%, which
was very close to a previously reported value of 67% (36). The
above reports, although limited in number, indicate that sweet
potato protein is of good nutritional quality but the quantity
is low in the cultivars used. The cultivar Tainon 57 used by

Huang et al. (35) had a crude protein content of from 0.8 to
1.3% (fresh weight).
        A report by Bressani et al. (37), which evaluated the
nutritional value of diets based on starchy foods and beans,
indicated that for the rat, sweet potato protein was of poor
nutritional quality. When methionine was added to all diets to
raise sulfur amino acids, sweet potato still required the largest
amount of supplementation with bean flour to maintain animal
weight (Table II).
        Sweet potato flour contained 3.8% protein, the second highest
amount of protein among starchy foods, and yet the protein
appeared to be the poorest in nutritional quality. However, it
should be noted that the sweet potatoes used in this study were
dried at 60°C but were not cooked. Uncooked sweet potato starch
is not completely digestable by rodents. As a consequence,
maintenance requirements would increase. This is the most likely
explanation for the increased requirement for bean flour, but
there also may have been interference with digestion from protease
inhibitors present in uncooked sweet potatoes.
        Walter et al. (38) measured the protein efficiency ratio
(PER) of flour prepared from sweet potatoes which were cooked
in a drying oven. Because the PER is determined on the basis
of a diet containing 10% protein, the 'Jewel' and 'Centennial'
sweet potatoes used in this study were stored until sufficient
starch had metabolized to increase crude protein content to 11.25%
(dry basis). When the flour was fed to Sprague-Dawley strain
rats, the corrected PER values were 2.22 and 2.00 for 'Centennial'
and 'Jewel' cultivars, respectively, compared to 2.50 for casein.
'Centennial' had the highest PER value of the two cultivars
because its NPN content was lower. The net effect of increased
NPN content is to lower the amount of essential amino acids as
a percentage of the total nitrogen and thus decrease the PER
value.

Anti-nutritional Factors

It has been recognized since 1954 (39) that sweet potato contains
trypsin inhibitors. Trypsin inhibitors (TI) have an anti-
nutritional effect by inhibiting proteolytic action of trypsin
during digestion. Since the initial report, TI activity in sweet
potatoes has been the subject of several reports. Dickey and
Collins (40) reported the presence of 7 TI bands in the 4
cultivars examined, the intensity of the bands being cultivar
dependent. Heat inactivation of TI also was cultivar dependent,
but heating the tissue to 94°C, followed by cooling to room
temperature destroyed 93-97% of the activity in all cultivars.
Consequently, cooking of sweet potatoes should eliminate most
of the anti-nutritional effect.
        Enteritis necrotians (EN), a spontaneous form of enteric
gangrene endemic to the highlands of Papua, New Guinea, is caused
by toxins produced when Clostridium perfringens of the gut enter
a rapid growth phase (41). It has been postulated that the disease
occurs in populations which consume a low protein diet, e.g.,
sweet potato as the staple food combined with TI activity which

Table II. Effect of Supplementation of Starchy Foods With Common Beans on Weight Maintenance[a]

Flours	% Crude Protein	% Bean Flour[b] Required for Nitrogen Balance
Cassava	1.4	14.5
Plantain	3.1	20.1
Potato	9.5	14.6
Sweet Potato	3.8	29.3
Bean	22.8	10.1[c]

[a] From Bressani et al. (37). Wistar rats were test animal.
[b] Supplemented with methionine.
[c] Cornstarch used as starchy food with bean flour.

effectively reduces the proteolytic capacity of the digestive
system to such a degree that it cannot destroy the proteinaceous
toxin by hydrolysis. A report by Bradbury et al. (13) indicated
that there was no correlation between the incidence of EN in
a given region and the amount of TI activity in the sweet potato
cultivars consumed in that region. Unless the populations involved
consume large amounts of raw sweet potatoes, it is highly unlikely
that the TI is obtained from this source since cooking has been
shown to inactivate the inhibitor (40, 42).

Amino Acid Composition

In recent years, a number of workers have published amino acid
analyses of the sweet potato (38, 43, 13, 22, 18). The overall
picture is that the sweet potato amino acid pattern is of good
nutritional quality but that the variability of individual amino
acids both within the same cultivar and across cultivars is very
high. For example, Walter et al. (44) reported that with the
exception of aromatic amino acids, every essential amino acid
has a score of less than 100 in one or more cultivars. The amino
acid score is defined as the g of amino acid in 100 g of test
protein divided by the number of g of that amino acid in the
FAO/WHO reference pattern times 100. Bradbury et al. (22) showed
that, for the same cultivar, environmental effects on the amino
acid patterns is significant. For three cultivars, they found
a mean percent standard deviation for all amino acids of 24.2,
23.4 and 20.6 over 5 environments. From their results, Bradbury
(22) concluded that in the highlands of Papua, New Guinea, the
EAA most likely to be limiting in decreasing order of probability
were lysine, leucine and sulfur amino acids. These workers
suggested that a part of the large difference reported worldwide
in the relative amount of sulfur amino acids may be due in part
to difficulties in the analysis of these compounds.

Concentrates and Isolates

The literature on concentrated sweet potato protein is sparse.
Amino acid patterns for sweet potato protein isolates have been
reported by three groups (16, 45, 46). One report showed that
when compared to the FAO standard (47), no amino acids were
limiting. The other reports showed total sulfur amino acids and
lysine to be limiting (Table III). The patterns indicate a
nutritionally well balanced protein. The improvement in
nutritional quality, when compared to amino acid patterns from
whole sweet potato, is due to the fact that whole sweet potatoes
contain substantial amounts of NPN, which consists mainly of
nonessential amino acids. This effectively dilutes the EAA and
lowers the amino acid score.
    Feeding studies with the rat as the test animal verified
the high nutritional quality indicated by the amino acid pattern
(45). Using isolates and concentrates prepared from 'Jewel' and
'Centennial' cultivars, PER values were equal to that of casein
(milk protein) (Table IV). Examination of the amino acid patterns
of sweet potato protein and casein revealed that both contained

Table III. Amino Acid Composition of Protein Isolates (g of Amino
Acid Per 100 g of Protein)

	Walter and Catignani (45)	Purcell et al. (16)[a]	Nagase (46)[b]	FAO/WHO (47)
Essential				
Threonine	6.4	5.5	4.6	4.0
Valine	7.9	6.8	7.9	5.0
Methionine	2.0	2.6	2.5	
Total Sulfur	3.1	3.0	4.1	3.5
Isoleucine	5.6	5.3	5.3	4.0
Leucine	7.4	7.8	8.7	7.0
Tyrosine	6.9	5.2	3.6	
Phenylalanine	8.2	6.7	6.0	
Lysine	5.2	6.8	6.5	5.5
Tryptophan	1.2[c]	1.1[c]	1.8[c]	1.0
Amino Acid Score[d,e]				
Total Sulfur	88	86	100	
Lysine	95	100	100	
Nonessential				
Aspartic Acid	18.9	14.4	13.1	
Serine	6.6	5.1	5.5	
Glutamic Acid	9.6	8.6	11.8	
Proline	4.2	5.4	4.3	
Glycine	5.3	4.3	2.6	
Alanine	5.4	4.6	6.1	
Histidine	2.7	2.4	4.2	
NH$_3$	1.6	-[f]	-[f]	
Arginine	5.9	6.0	6.4	

[a]'Jewel' cultivar.
[b]Cultivar unknown.
[c]Tryptophan content measured colorimetrically on enzyme-hydrolyzed material.
[d]g of amino acid in 100 g of test protein/g of amino acid in FAO/WHO reference pattern x 100.
[e]All other essential amino acids exceeded FAO/WHO values.
[f]NH$_3$ not reported.
From Walter et al. (44).

Table IV. Protein Efficiency Ratio (PER) [a] for Protein Fractions From Sweet Potatoes

Protein Fractions	PER	Corrected PER	Wt. Gained, g	Food Consumed, g	Initial Group wt., g
White					
Casein	2.81 ± 0.11	2.50 ± 0.09	134.3 ± 11.7	477.9 ± 37.7	78.3 ± 3.1
'Jewel'	2.91 ± 0.10	2.64 ± 0.09	138.9 ± 11.7	477.1 ± 29.0	78.3 ± 3.3
'Centennial'	2.96 ± 0.07	2.63 ± 0.07	140.3 ± 12.4	472.6 ± 35.3	78.4 ± 3.2
Chromoplast					
Casein	2.78 ± 0.10	2.50 ± 0.09	109.5 ± 7.8	394.0 ± 25.3	71.6 ± 2.9
'Jewel'	2.73 ± 0.09	2.47 ± 0.09	117.6 ± 11.3	431.1 ± 39.5	71.1 ± 2.7
'Centennial'	2.78 ± 0.10	2.50 ± 0.10	122.2 ± 14.9	437.9 ± 44.5	71.3 ± 2.7

[a] Mean and standard deviation calculated from data from 10 rats per diet group.
[b] Corrected by adjusting test diets to 2.50 for casein (AOAC).
From Walter and Catignani (45).

less sulfur amino acids than required for rat growth. In addition, sweet potato contained less lysine, while casein contained less threonine than is required for rat growth. Apparently the overall deficiencies limited rat growth about the same amount. The end result was that rats fed either protein grew at about the same rate.

Horigome et al. (15) reported a PER of 1.9 for protein recovered from an industrial sweet potato starch facility. They were able to increase the PER to 2.5 by supplementing the diets with lysine and methionine. A portion of these amino acids were either destroyed or made biologically nonavailable by the processing operation. The possibility also exists that these amino acids were limiting in the cultivars studied.

Effect of Processing on Nutritional Quality

Heat processing of sweet potatoes can have deleterious effects on protein nutritional quality. Purcell and Walter (48) found that the intensity of the heat processing conditions had a direct bearing on nutritional quality of the protein. In this study lysine was destroyed, presumably via irreversible reaction with reducing sugars (40). Both sucrose syrup-canned sweet potatoes and drum-dried sweet potato flakes contained 26% less lysine than did baked sweet potatoes. In addition, syrup-canned sweet potatoes contained 25% less total nitrogen than did either baked or drum-dried sweet potatoes. This loss of nitrogen was apparently due to solution of the NPN fraction in the syrup. Other reports on canned sweet potatoes reveal similar changes. Canned sweet potatoes from various locations were found to contain 3.8 to 4.2% (dry basis) crude protein (50), rather than the expected 4.5-7.0%. Although no mention was made of the lower-than-expected crude protein values, these were probably due to dissolution of part of the NPN fraction in the syrup. Similarly, Meredith and Dull (43) reported that canned-in-syrup sweet potatoes contained ca. 45% less amino acids than did the roots before processing. Since syrup is discarded before the canned roots are eaten, this results in a serious loss of nitrogen.

The severity of heat treatment during dehydration has a significant effect on protein nutritional quality. Cooked sweet potatoes dehydrated in a forced-draft oven at 60°C had a PER of 2.2, while a second lot of cooked sweet potatoes dehydrated on a steam-heated drum dryer had a PER of 1.3 (38). The lysine content measured by acid hydrolysis-ion exchange chromatography was somewhat lower in the drum dehydrated flour but not sufficiently low to account for the difference in PER values. Further study using an assay for available lysine (51) showed that a large part of the lysine was not available. Thus, acid hydrolysis can liberate biologically nonavailable lysine which is subsequently quantified along with available lysine, causing an overestimation of the nutritional quality of the food. This is most likely to happen when high levels of reducing sugars are present in the food and lysine is limiting, as is the case with sweet potatoes.

Summary and Conclusions

The sweet potato ranks sixth in average production among the major food crops of the world. There is significant potential for increasing the protein content of this crop by a combination of breeding/selection and optimization of production practices. According to present knowledge, most of the nitrogen of the sweet potato is in a form suitable to satisfy human nitrogen requirements. The protein component comprises from 60-85% of the nitrogen with the remainder consisting of amino or amide nitrogen. The amino acid pattern of the sweet potato is highly variable. Isolated sweet potato protein is of sufficient nutritional quality to support growth of laboratory rats to the same extent as casein. Humans have been maintained in nitrogen balance using sweet potato as the major source of protein. Processing of sweet potatoes can have adverse effects on the protein nutritional value. Canning sweet potatoes in a liquid medium causes leaching of soluble nitrogenous compounds into the liquid, thereby lowering the nitrogen content. Heat processing of the sweet potato causes a decrease in the biological availability of lysine. The extent of the decrease in lysine availability is dependent upon the severity of the heat treatment and the amount of reducing sugars present during heating.

Acknowledgments

Paper no. 10141 of the Journal Series of the North Carolina Agricultural Research Service, Raleigh, NC 27695-7601. Mention of a trademark or proprietary product does not constitute a guarantee or warranty of the product by the U. S. Department of Agriculture or North Carolina Agricultural Research Service, nor does it imply approval to the exclusion of other products that may be suitable.

Literature Cited

1. "Production Yearbook," FAO, 1977, Rome, Italy.
2. "Recommended Daily Allowances," Food and Nutrition Board, National Academy of Sciences, National Research Council, 1980, Washington, DC.
3. Watt, B. K.; Merrill, A. L. "Composition of Foods," 1975, U. S. Department of Agriculture Handbook No. 8.
4. USDA. Economics and Statistics Service, 1980, Statistical Bulletin No. 645.
5. Hipsley, E. H.; Kirk, N. E. Technical paper, 1965, South Pacific Commission, New Calonia, No. 147.
6. Purcell, A. E.; Walter, W. M., Jr. J. Agric. Food Chem. 1980, 28, 842.
7. Dickey, L. F.; Collins, W. W.; Young, C. T.; Walter, W. M., Jr. Hortscience, 1984, 19, 689.
8. Purcell, A. E.; Walter, W. M., Jr.; Giesbrecht, F. G. J. Amer. Soc. Hort. Sci. 1978, 103, 190.
9. Sober, H. A. "CRO Handbook of Biochemistry. Selected Data for Molecular Biology"; Chemical Rubber Co.: Cleveland, OH, 1970; pp. 1394-1395.

10. Jones, D. B.; Gersdorff, C. E. F. J. Biol. Chem. 1931, 93, 119.
11. Li, He-S.; Oba, K. Agric. Biol. Chem. 1985, 49, 737.
12. Purcell, A. E.; Walter, W. M., Jr.; Giesbrecht, F. G. J. Agric. Food Chem. 1976, 24, 64.
13. Bradbury, J. H.; Baines, J.; Hammer, B.; Anders, M.; Millar, J. S. J. Agric. Food Chem. 1984, 32, 469.
14. Purcell, A. E.; Walter, W. M., Jr.; Giesbrecht, F. G. J. Agric. Food Chem. 1978, 26, 699.
15. Horigome, T.; Nakayama, N.; Ikeda, M. Chem. Abstr. 1972, 77, 661N.
16. Purcell, A. E.; Swaisgood, H. E.; Pope, D. T. J. Amer. Soc. Hort. Sci. 1972, 97, 30.
17. Li, L. J. Agric. Assoc. China 1974, 88, 17.
18. Goodbody, S. Trop. Agric. (Trinidad) 1984, 61, 20.
19. Li, L. J. Agric. Assoc. China 1977, 100, 78.
20. Purcell, A. E.; Walter, W. M., Jr.; Giesbrecht, F. G. J. Agric. Food Chem. 1978, 26, 362.
21. Collins, W. W.; Walter, W. M., Jr. "Sweet Potato: Proceedings of the First International Symposium"; Villareal, R. L.; Griggs, T. D., eds., Asian Vegetable Research and Development Center, Shanhua, Taiwan, China, 1982, p. 355.
22. Bradbury, J. H.; Hammer, B.; Hguyen, T.; Anders, M.; Miller, J. S. J. Agric. Food Chem. 1985, 33, 281.
23. Purcell, A. E.; Walter, W. M., Jr.; Nicholaides, J. J.; Collins, W. W.; Chancy, H. J. Amer. Soc. Hort. Sci. 1922, 107, 425.
24. Constantin, R. J.; Jones, L. G.; Hammett, H. L.; Hernandez, T. P.; Kahlich, C. G. J. Amer. Soc. Hort. Sci. 1984, 105, 610.
25. Kimber, A. J. Papua New Guinea Food Crops Conference Proceedings, Dept. Prim. Indus., Wilson, K.; Bourke, R. M., eds., Port Moresby, New Guinea, 1975, p. 63.
26. Li, L. J. Agric. Assoc. China 1975, 92.
27. Yeh, T. P.; Chen, Y. T.; Sun, C. C. J. Agric. Assoc. China 1981, 113, 33.
28. Purcell, A. E.; Pope, D. T.; Walter, W. M., Jr. Hortscience, 1976, 11, 31.
29. Constantine, R. J.; Hernandez, T. P.; Jones, L. G. J. Amer. Soc. Hort. Sci. 1974, 99, 308.
30. Knorr, D. Lebensm. -Wiss. Technol. 1978, 11, 109.
31. Adolph, W. H.; Liu, H. C. Chin. Med. J. 1939, 55, 337.
32. Kao, H. C.; Adolph, W. H.; Liu, H. C. Chin. J. Physiol. 1935, 9, 141.
33. Ruinard, J. Proc. Int. Symp. Tropical Crops 1967, 1, 89.
34. Oomen, H. A. P. C. Proc. Nutr. Soc. 1970, 29, 197.
35. Huang, P. C.; Lee, N. Y.; Chen, S. H. Amer. J. Clin. Nutr. 1979, 32, 1741.
36. Kandatsu, M. In "Food Chemistry"; Koseikan, Tokyo, 1964, p. 108.
37. Bressani, R.; Navarrete, D. A.; Elias, L. G. Qual. Plant Plant Foods Human Nutr. 1984, 34, 109.
38. Walter, W. M., Jr.; Catignani, G. L.; Yow, L. L.; Porter, D. H. J. Agric. Food Chem. 1983, 31, 947.

39. Sohonie, K.; Bhandarker, A. P. J. Sci. Ind. Res. 1954, 13B, 500.
40. Dickey, L. F.; Collins, W. W. J. Amer. Soc. Hort. Sci. 1984, 109, 750.
41. Murrell, T. G. C. Chin. Med. J. 1982, 95, 843.
42. Obidairo, T. K.; Akpochago, O. M. Enzyme Microbiol. Technol. 1984, 6, 132.
43. Meredith, F. I.; Dull, G. G. Food Technol. 1979, 33, 55.
44. Walter, W. M., Jr.; Collins, W. W.; Purcell, A. E. J. Agric. Food Chem. 1984, 32, 695.
45. Walter, W. M., Jr.; Catignani, G. L. J. Agric. Food Chem. 1981, 29, 797.
46. Nagase, T. Fukuoka Igaku Zasshi, 1957, 48, 1828.
47. FAO/WHO. W.H.O. Tech. Rep. Ser. 1973, 522.
48. Purcell, A. E.; Walter, W. M., Jr. J. Agric. Food Chem. 1982, 30, 443.
49. Carpenter, K. J. Nutr. Abstr. Rev. 1973, 43, 404.
50. Collins, J. L. Tenn. Farm Home Sci. 1981, Jan.-Mar., 25.
51. Goodno, C. C.; Swaisgood, H. E.; Catignani, G. L. Anal. Biochem. 1981, 115, 203.

RECEIVED December 26, 1985

Cucurbit Seed Protein and Oil

T. J. Jacks

Southern Regional Research Center, Agricultural Research Service, U.S. Department of Agriculture, New Orleans, LA 70179

Research concerning the structure, composition, and usefulness of cucurbit seeds (gourds, melons, squash, etc.) is reviewed. Cytological features are typical of those for oilseeds. Composition-ally, decorticated seeds contain by weight 50% oil and 35% protein. The oil is unsaturated and edible; however, certain species contain conju-ated trienoic fatty acids (drying oils). Globulins account for 70 to 90% of the protein and consist of two, four or six subunits of 54,000 daltons. Disulfide reduction of the subunit yields polypeptides of 19,000 to 37,000 daltons. Globulins are rich in arginine, aspartic and glutamic acids, and are deficient in lysine and sulfur-containing amino acids. Nutritional values of the globulin are similar to those of other oilseed globulins; supple-mentation with the limiting amino acids increases the values.

The first general review of the composition and characteristics of oil and protein from several species of cucurbit seeds was published in 1972 (1). Since then, a resurgence of interest has developed in the exploitation of seeds from wild, xerophytic cucurbits, especially Buffalo gourd (Cucurbita foetidissima), as useful, nutritious food-stuff components (2-4). Reviews concerning results of field studies of Buffalo gourd production and a description of the University of Arizona's program to domesticate it have also appeared (5,6). In addition, seeds of other cucurbits seem just as economically and nutritionally promising. In this chapter, pertinent results of research on cucurbit seeds with regard to yield, cytological struc-ture, composition, and characterizations and nutritional aspects of oil and protein are summarized.

Yield

Seed yield in cucurbits is seldom investigated because commercial production of the carbohydrate-rich fruit is the major concern. Wide variations in sizes and weights of seeds, numbers of seeds per fruit and even numbers of fruits per plant seems the rule, particularly in wild plants and even within one species (7). Estimations from limited observations of C. foetidissima, C. digitata and C. palmata growing wild in desert areas indicate theoretical yields from 500 to 3,000 lb of seeds per acre (8,9). C. foetidissima cultivated in northwestern Texas yields approximately 700 to 2,000 lb of seeds per acre (10). C. pepo (pumpkin) produces up to 1,200 lb per acre and an improved seed-coatless line yields from 1,200 to 1,400 lb of seed per acre (9, 11). These yields are comparable to yields of oilseeds of commerce.

Cytological Structure

Cucurbit seeds are exalbuminous or lacking endosperm in the mature state. In such seeds the embryo is large in relation to the seed as a whole. It fills the seed almost completely and its body parts, particularly the cotyledons, store the food reserves for germination. Since the predominant tissue of the seed is cotyledonous, and since cotyledons are leaves, anatomy and histology of typical leaf tissue suffice to describe the preponderant part of the seed. Epidermal cells cover the cotyledonary surface followed by palisade and abundant parenchyma cells that contain the food reserves. Vascular tissues are also present.

A cross-sectional view of a C. digitata seed is shown in a scanning electron micrograph (SEM) in Figure 1. The section was treated with hexane before being sputter-coated (12) and is morphologically identical to a similar view of C. pepo (12). The seed coat comprises the somewhat thin outer boundary of the section and the remainder is composed of two cotyledons separated by the first "true" leaves of the embryo.

To show the intracellular structure of a cucurbit seed, this time C. foetidissima, an SEM at higher magnification is given in Figure 2. Within the cell wall are large particles of protein (protein bodies) and a cytoplasmic reticulum in which oil-rich spherosomes were embedded. This emptied spherosoma complex, appearing as a net-work, is preserved (12, 13) when the sample is prepared by the aqueous method of Arnott and Webb (12). When the method given for Figure 1 was used, the intracellular structure (protein bodies but no cytoplasmic reticulum) was similar to that of C. foetidissima shown by Tu et al (13).

Cytological features of the cotyledons are shown in Figure 3, which is a composite electron micrograph portraying typical paren-chyma cells that comprise the cotyledonary storage tissues of C. foetidissima, C. pepo, C. palmata, C. digitata, and Apodanthera undulata (14). The bulk of the cytoplasm consists of two organelles: spherosomes (lipid bodies) and protein bodies (aleurone grains). Starch grains are absent (1). Spherosomes are about 1 micron in diameter, are surrounded by half-unit membranes (15), and contain the reserve oil of oilseeds (16). Protein bodies are from 5 to 20 microns in diameter, are enclosed in unit membranes, contain storage protein (17, 18), and harbour two inclusions: crystalloids and globoids. Crystalloids are crystalline deposits of storage protein

Figure 1. Scanning electron micrograph of a cross section of a dormant, hexane-treated seed of Cucurbita digitata. Note outer seed coats, two cotyledons (C), and central cleft that contains first "true" leaves between the cotyledons. Bar represents 1 mm.

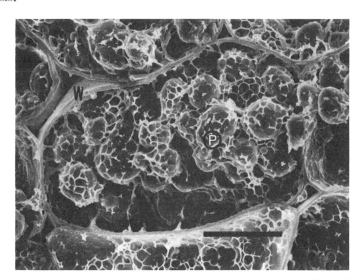

Figure 2. Scanning electron micrograph of a mesophyll cell of a dormant cotyledon of Buffalo gourd (Cucurbita foetidissima). Tissue was fixed in aqueous glutaraldehyde, dehydrated with ethanol and critically point dried. Note cell wall (W) and intracellular components including protein bodies (P) and emptied spherosomes that appear as a cytoplasmic reticulum. Bar represents 10 μm.

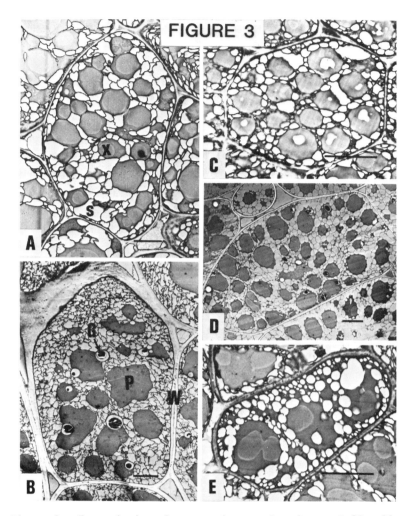

Figure 3. Transmission electron micrographs of mesophyll cells of dormant cotyledons of: A, Cucurbita foetidissima; B, Cucurbita pepo; C, Cucurbita palmata; D, Cucurbita digitata; E, Apodanthera undulata. Note cell wall (W), protein body (P), spherosome (S), globoid (G), and crystalloid (X). In each micrograph, the bar represents five microns. Reproduced from reference 14.

(cucurbitin) and are generally abundant and large in cucurbits (17, 18). Globoids are composed mostly of metallic salts of phytic acid (19, 20). Energy dispersive x-ray analysis of C. maxima (squash) globoids have shown that, in addition to phosphorus due to phytic acid, potassium, magnesium, and sometimes calcium are located in globoids (17). These cations comprise the metallic salts of phytin in the globoid; they are absent in the proteinaceous matrix of the protein body in which globoids are embedded (17, 19).

Other intracellular organelles, such as mitochondria, plastids, and endoplasmic reticula, all of which are rarely observed in the cytoplasm of quiescent seed cells, are not apparent in quiescent cucurbit seed cotyledons. Nuclei, however, are present.

Composition

As with yields of seeds given above, the amount of seed coat per seed varies considerably, anywhere from 18% (C. pepo) to 60% (Lagenaria vulgaris, bottle gourd) (21, 22). Indeed, seed-coatless lines of C. pepo have little or no coat (11, 23).

The amounts of oil and protein in decorticated seeds are somewhat less varied. Calculations of earlier data from thirteen species show that decorticated seeds contain, by weight, 49.5 + 2.3% oil and 35.0 + 2.4% protein at 95% confidence intervals in the Z test (1). More recent reports (7, 24-32) are in substantial agreement with these values. Some studies report oil and protein contents of undecorticated seed (whole seed) or protein content of oil-free meal. Recalculations to decorticated full-fat seeds fit these ranges.

Oil

Unsaturated fatty acids are the preponderant fatty acids of cucurbit oils, and in some seeds conjugated triene comprises one-third of this unsaturation. Table I shows the fatty acid distribution in oils of cucurbit seeds (1). More recent determinations (7, 27, 31, 33) are in close agreement with these results. Occasionally a species

Table I. Content of Fatty Acids in Cucurbit Oils*

	Palmitic	Stearic	Oleic	Linoleic	Conjugated Trienes
Edible Oils	14.2 + 3.1	8.4 + 2.5	28.5 + 4.1	47.3 + 4.5	
Drying Oils	7.8 + 2.9	8.4 + 5.9	22.4 + 5.7	31.0 + 8.4	29.2 + 6.7

*Values are means of the percentages of weights + the limits of the 95% confidence intervals calculated from the Z test. Data from more than one determination on a given species were averaged before the means of data from all species were calculated so that each species was evaluated, or weighted, equally. Reproduced from reference 1.

contains an unusual amount of a given fatty acid, such as 49% oleic acid in seed-coatless C. pepo (26) or 78% linoleic acid Lagenaria masacarena (34). Clearly, the data show that cucurbit seeds are important sources of edible oil, and digestibility studies with

chicks and rats supports this (27, 33, 35), unless conjugated trienes
are present. In that case, the oil is valuable as a drying oil and
studies of C. digitata and C. palmata seed oils illustrate their
usefulness as protective coatings (8).

It is of interest that carotenoid pigments (xanthophylls),
sterols (spinasterol and chondrillasterol) and a triterpene alcohol
have been identified in cucurbit seed oil (31, 36, 37). However,
cucurbit oils such as that from Buffalo gourd are amenable to
refining, bleaching and deodorizing (38).

Protein

Decorticated cucurbit seeds contain by weight about 35% protein.
Traditionally, seed proteins are classified as globulins and albumins
according to their solubility in certain aqueous solvents.
Biochemically, oilseed globulins are generally considered storage
proteins while albumins are believed to be metabolic (catalytic)
proteins.

Albumins have not been as thoroughly investigated as have
globulins. Albumins isolated from Citrullus vulgaris (watermelon)
and C. maxima are composed of 9-12 major components that differ in
electrophoretic migration (39-41) and 6-9 components in gel
filtration (40). Many electrophoretically distinguishable proteins
comprise the albumin fraction of Cucumis sativus (cucumber), which is
dominated by albumins of molecular weight of 7000 to 9000 daltons
(42). A thorough study of Cucumis sativus albumin (43) showed that
about one-fourth of its protein is water soluble. This low molecular
weight albumin has a sedimentation constant of 2 S (Svedberg units)
in the analytical ultracentrifuge. It was concluded that, besides
cucurbit globulins, a large portion of the albumins also acts
biochemically as storage protein, but for sulfur in addition to
nitrogen since their amino acid composition is similar to that of
nitrogen-rich globulin, yet they contain an exceptionally high
content of cysteine (8.9% of the total amino acids).

Cucurbit seed globulins, which account for about 70 to 90% of
the total protein content, contain about 18% nitrogen, are soluble in
10% salt solutions from which they readily crystallize upon dilution,
and are also soluble in both acidic and basic solutions of low ionic
strength. Since the classical isolation of crystalline cucurbitin in
1892 (44), many chemical and physicochemical studies of cucurbitin
have been conducted. Varied results from determinations of the
molecular weight of the native oligomeric globulins and the number
and molecular weights of its subunits have been obtained. In
addition to slight natural variations among species, the variability
appears due to preparative procedures as well as to conditions and
modes of analyses. Earlier studies of the molecular weight of the
oligomeric globulin, the number of subunits and their electrophoretic
heterogeneity have been reviewed (1). Results prior to 1972
indicated that cucurbitin is a hexamer of about 340,000 molecular
weight and sedimentation constants in three states of association-
disassociation are 3 S (monomeric subunit), 7 S and 12 S. At acidic
pH values or high ionic strengths, cucurbitin appears homogeneous
during electrophoresis or ultracentrifugation; however, at neutral pH
values and low ionic strengths, the protein contains up to four major
components.

A later study (45) indicates that cucurbitin from pumpkin has a molecular weight of 112,000 daltons that can be electrophoretically separated into subunits of 63,000 and 56,000 daltons. Reduction of disulfides produces polypeptides of 36,000 and 22,000 daltons. Globulins from six cucurbits examined chromatographically (46) have molecular weights of 220,000 to 260,000 daltons that exhibit predominantly 10.4 - 11.2 S values (about 95% of the three globulin fractions). Cucurbitin from Cucumis sativus appears a tetramer of 240,000 daltons composed of four subunits of 54,000 daltons (42). Disulfide reduction produces six polypeptide subunits ranging from 37,000 to 19,000 daltons. A more recent study of cucurbitin isolated from several species (47) indicates that it is composed of only 12 S (90% of the globulin) and 18 S values. By electrophoretic analysis and chromatography combined with ultracentrifugation, native cucurbitin appears a hexamer of about 325,000 daltons (12 S form), composed of six monomeric subunits of about 54,000 daltons each, and forms a dimer of 630,000 daltons (18 S form). The large and small polypeptide subunits of the monomer, prepared by reduction with 2-mercaptoethanol, range from 33,000 to 36,000 daltons and 22,000 to 25,000 daltons, respectively.

Studies of the secondary structure of cucurbitin have shown its conformational modes consist of 5% α-helical, 32% pleated sheet, and 62% unordered structures (48). These values are similar in distribution to those of other oilseed globulins (48).

The amino acid compositions of total cucurbit seed protein (meal) and purified globulin are presented in Table II. More recent compositional data (7, 29, 32, 43, 45, 46, 49-51) since those summarized earlier (1) are in substantial agreement with the mean distribution shown in Table II, which indicates that cucurbit seeds, as oilseeds in general, are rich in glutamic acid (and glutamine), arginine, and aspartic acid (and asparagine). The abundance of these nitrogen-rich amino acids accounts for the 18% nitrogen content of cucurbit protein. Other nonprotein amino acids have been identified in cucurbit seeds (52), including 3-amino-3-carboxypyrolidine (cucurbitine), which has anthelmintic activity (53).

From amino acid compositions, evaluations of the nutritional potentials of cucurbit meals and globulins can be calculated according to FAO/WHO (54). The A:E ratios, which are the amounts of each essential amino acid relative to the total amount of essential amino acids, are shown in Table II. These data indicate that, like most other oilseeds, cucurbit seeds are deficient in lysine and sulfur-containing amino acids. However, sulfur-containing amino acids are considerably high in Citrullus colocynthis (egusi, ancestral watermelon) seed protein and exceed the suggested level in FAO/WHO reference protein (55).

A protein that is unduly rich in the ten essential amino acids would not provide sufficient nitrogen for other metabolic processes without obligatory catabolism of the essential amino acids. Thus, the proportion of the total nitrogen intake that essential amino acids form indicate how a given protein fulfills nutritional requirements for proteins. This proportion, the E/T ratio (54), indicative of the amount of protein nitrogen supplied by essential amino acids, is (in g of essential amino acids per g of nitrogen) 2.18 for cucurbit meal and 2.67 for cucurbit globulin. The value for meal is similar in magnitude to those for other seeds and the value

Table II. Amino Acid Contents[a] and A/E Ratios[b] of Cucurbit Meals and Globulins

Amino Acid	Meal Content	Meal A/E Ratio	Globulin Content	Globulin A/E Ratio	FAO Pattern[c] Content	FAO A/E Ratio
Cysteine	1.0	29	1.5 ± 0.3	43	2.0[d]	62[d]
Isoleucine	4.3 ± 0.5	124	4.8 ± 0.1	112	4.2	134
Leucine	6.5 ± 0.7	187	8.3 ± 0.3	194	4.8	152
Lysine	4.3 ± 1.4	124	3.0 ± 0.2	70	4.2	134
Methionine	1.8 ± 0.4	52	3.0 ± 0.2	66	2.2[d]	71[d]
Phenylalanine	4.4 ± 0.4	126	7.3 ± 0.5	171	2.8	89
Threonine	2.7 ± 0.4	78	3.2 ± 0.3	75	2.8	89
Tryptophan	1.6 ± 0.1	46	2.1 ± 0.2	49	1.4[d]	15[d]
Tyrosine	3.0 ± 0.1	86	3.8 ± 0.3	89	2.8	89
Valine	5.2 ± 1.3	149	5.9 ± 0.3	138	4.2	134
Alanine	4.2		5.2 ± 0.3			
Arginine	13.8 ± 1.3		16.6 ± 0.6			
Aspartic Acid	9.0 ± 1.9		9.8 ± 1.1			
Glutamic Acid	16.1 ± 2.4		20.0 ± 1.7			
Glycine	5.4 ± 1.3		5.1 ± 0.2			
Histidine	2.3 ± 0.5		2.5 ± 0.2			
Proline	3.3		5.1 ± 0.2			
Serine	4.6 ± 0.3		5.9 ± 0.7			

a Values are the means of g of amino acid per 16 g of nitrogen ± the limits of the 95% confidence intervals calculated from the Z test. Limits are not shown where less than three values were averaged. Data from more than one determination on a given species were averaged before calculations of data from all the species so that each species was weighted equally.
b Values are mg of amino acid per g of total essential acids (18).
c From reference 18.
d Probably too high (18).
Reproduced from reference 1.

for globulin is similar to those for fish and pork tenderloin (54).
The nutritional properties of cucurbit seeds have also been evaluated
experimentally through feeding studies with animals. Results
summarized in 1972 (1) show that cucurbit seed products are inferior
to milk products for supporting weight increases in growing rats, but
that supplementation of cucurbit meals with certain limiting amino
acids increases the nutritional value of the meals.

Recent nutritional evaluations of cucurbit seeds have
accentuated the currently popular Buffalo gourd (C. foetidissima) and
are shown in Table III. PER (protein efficiency ratio) values of
full-fat and natural and autoclaved fat-free meals fed to weanling

Table III. Nutritional Evaluations of Buffalo Gourd

Value	Egg/Casein	Meal	Globulin
E/T		2.18	2.67
	6.9a	3.3-4.7	
	2.5b		1.7
NPR	4.0a	2.2	
Digestibility (%)	82a	70	93
aEgg			
bCasein			

mice range from 2.0 to 2.8 for C. digitata and 3.3 to 4.7 for C.
foetidissima (51). Protein digestibilities are about 70%. The
meals are inferior to whole egg (6.9 PER; 82% digestibility). In a
separate study (32), the feeding of full-fat C. digitata, C.
foetidissima, and Apodanthera undulata meals to weanling mice yielded
PER values of 0.4, 1.5 and 1.2, respectively, compared to 2.5 for
whole egg. One-hundred percent mortality within five days results
from ingestion of Citrullus colocynthis (egusi). The NPR values (net
protein retention) are 1.8, 2.2 and 2.4, respectively, compared to
4.0 for whole egg. Values greater than 2.0 are considered good
protein quality. Fat-free C. foetidissima meal fed to growing rats
is similar in nutritive value to soybean meal, and upon supplementation with lysine and threonine, approaches casein (56). When
crystallized C. foetidissima globulin is the dietary source of
protein for weanling rats, the corrected PER is 1.7 compared to 2.5
for casein, and digestibility is 93% (49). However, a somewhat weak
trypsin-inhibitor activity is present in C. foetidissima seeds (28).

Carbohydrates

Ubiquitous in seeds is phytic acid, the hexaphosphate ester of
inositol, which has been isolated from cucurbit seeds (57). Small
amounts of free sugars and terpenoid glycosides (cucurbitacins) are
also present (58-60), but starch is absent (1, 61). Cellulosic cell
wall materials comprise the remaining carbohydrate content.

Literature Cited

1. Jacks, T. J.; Hensarling, T. P.; Yatsu, L. Y. Econ. Bot. 1972, 26, 135-41.
2. Bemis, W. P.; Curtis, L. C.; Weber, C. W.; Berry, J. W.; Nelson, J. M. Agency for Internat. Devel. Tech. Series Bull. 1975, 20 p.
3. Hinman, C. W. Science 1984, 225, 1445-48.
4. Nat. Acad. Sci. "Underexploited Tropical Plants With Promising Economic Value"; National Academy of Sciences: Washington, D.C., 1975; pp. 94-99.
5. Bemis, W. P.; Berry, J. W.; Weber, C. W. In "New Agricultural Crops"; Ritchie, G. A., Ed.; AAAS Selected Symposium No. 38; Westview Press: Boulder, CO, 1979; pp. 65-87.
6. Vasconcellos, J. A.; Bemis, W. P.; Berry, J. W.; Weber, C. W. In "New Sources of Fats and Oils"; Pryde, E. H.; Princen, L. H.; Mukherjee, K. D., Eds.; Am. Oil Chem. Soc.: Champaign, IL, 1981; pp. 55-68.
7. Scheerens, J. C.; Bemis, W. P.; Dreher, M. L.; Berry, J. W. J. Am. Oil Chem. Soc. 1978, 55, 523-525.
8. Bolley, D. S.; McCormack, R. H.; Curtis, L. C. J. Am. Oil Chem. Soc. 1950, 27, 571-574.
9. Curtis, L. C. Chemurgic Dig. 1946, 13, 221-224.
10. Shahani, H. S.; Dollear, F. G.; Markley, K. S.; Quimby, J. R. J. Am. Oil Chem. Soc. 1951, 28, 90-95.
11. Curtis, L. C. Proc. Am. Soc. Hort. Sci. 1948, 52, 223-224.
12. Arnott, H. J.; Webb, M. A. In "New Frontiers in Food Microstructure"; Bechtel, D. B., Ed.; Am. Assoc. Cereal Chem.: St. Paul, MN, 1983; pp. 149-198.
13. Tu, M.; Deyoe, C. W.; Eustace, W. D. Cereal Chem. 1978, 55, 773-778.
14. Hensarling, T. P.; Jacks, T. J.; Yatsu, L. Y. J. Am. Oil Chem. Soc. 1974, 51, 474-475.
15. Yatsu, L. Y.; Jacks, T. J. Plant Physiol. 1972, 49, 937-943.
16. Yatsu, L. Y.; Jacks, T. J.; Hensarling, T. P. Plant Physiol. 1971, 48, 675-682.
17. Lott, J. N. A. In "The Biochemistry of Plants. A Comprehensive Treatise. Vol 7. The Plant Cell"; Tolbert, N. E., Ed.", Academic Press: New York, NY, 1980; pp. 589-623.
18. Pernollet, J. C. Phytochemistry 1978, 17, 1473-80.
19. Lui, N. S. T.; Altschul, A. M. Arch. Biochem. Biophys. 1967, 121, 678-684.
20. Wada, T.; Maeda, E. Japan. J. Crop Sci. 1980, 49, 51-57.
21. Arevalo, J. J. Studia Rev. Univ. Atlantico, Barranquilla 1957, 2, 138-209.
22. Chowdhury, D. K.; Chakrabarty, M. M.; Mukherji, B. K. J. Am. Oil Chem. Soc. 1955, 32, 384-386.
23. Jaky, M. Oleagineux 1958, 13, 149-151.
24. Bemis, W. P.; Scheerens, J. C.; Berry, J. W.; Dreher, M. L.; Weber, C. W. J. Am. Oil Chem. Soc. 1977, 54, 537-538.
25. Berry, J. W.; Weber, C. W.; Dreher, M. L.; Bemis, W. P. J. Food Sci. 1976, 41, 465-466.
26. Black, J. M.; Rhew, T. H.; Melton, S. L. J. Am. Oil Chem. Soc. 1980, 57, 136A.
27. Khoury, N. N.; Daghir, S.; Sawaya, W. J. Food Technol. 1982, 17, 19-26.

28. Lancaster, M.; Storey, R.; Bower, N. W. Econ. Bot. 1983, 37, 306-309.
29. Robinson, R. G. Agron. J. 1975, 67, 541-544.
30. Tu, M.: Deyoe, C. W. Cereal Sci. Today 1974, 19, 404.
31. Vasconcellos, J. A.; Berry, J. W.; Weber, C. W.; Bemis, W. P.; Scheerens, J. C. J. Am. Oil Chem. Soc. 1980, 57, 310-313.
32. Weber, C. W.; Berry, J. W.; Philip, T. Food Tech. 1977, 31, 182-183.
33. Sawaya, W. N.; Daghir, N. J.; Khan, P. K. J. Food Sci. 1983, 48, 104-110.
34. Gunstone, F. D.; Taylor, G. M.; Cornelius, J. A.; Hammonds, T. W. J. Sci Fd. Agric. 1968, 19, 706-709.
35. Asenjo, J. J.; Goyco, J. A. Bol. Colegio quim. Puerto Rico 1951, 8, 14-16.
36. Iida, T.; Jeong, T. M.; Tamura, T.; Matsumoto, T. Lipids 1980, 15, 66-68.
37. Itoh, T.; Jeong, T. M.; Tamura, T.; Matsumoto, T. Lipids 1980, 15, 122-123.
38. Vasconcellos, J. A.; Berry, J. W. J. Am. Oil Chem. Soc. 1982, 59, 79-84.
39. Kretovich, V. L.; Bundel, A. A.; Lepina, N. A.; Melik-Sarkisyan, S. S. Biokhim. Zerna, Sbornik 1956, 5-34; Chem. Abstr. 1957, 51, 13970g.
40. Mourgne, M.; Barbe, J.; Campenio, S.; Lanet, J.; Savary, J.; Vinet, L. Ann. Pharm. Fr. 1971, 20, 583-590.
41. Mourgne, M.; Baret, R.; Renai, J.; Barbe, J. C. R. Soc. Biol. 1966, 160, 1166-69.
42. Koller, W.; Frevert, J.; Kindl, H. Hoppe-Seyler's Z. Physiol. Chem. 1979, 360, 167-176.
43. Youle, R. J.; Huang, A. H. C. Am. J. Bot. 1981, 68, 44-48.
44. Osborne, T. B. Am. Chem. J. 1892, 14, 662-689.
45. Hara, I.; Wada, K.; Wakabayashi, S.; Matsubara, H. Plant Cell Physiol. 1976, 17, 799-814.
46. Pichl, I. Phytochemistry 1976, 15, 717-722.
47. Blagrove, R. J.; Lilley, G. G. Eur. J. Biochem. 1980, 103, 577-584.
48. Jacks, T. J.; Barker, R. H.; Weigang, O. E. Int. J. Peptide Protein Res. 1973, 5, 289-291.
49. Hensarling, T. P.; Jacks, T. J.; Booth, A. N. Agric. Food Chem. 1973, 21, 986-988.
50. Mourgne, M.; Barbe, J.; Campenio, S.; Lanet, J.; Savary, J.; Vinet, L. Ann. Pharm. Fr. 1971, 29, 583-590
51. Weber, C. M.; Bemis, W. P.; Berry, J. W.; Deutschman, A. J.; Reid, B. L. Soc. Exper. Biol. Med. 1969, 130, 761-765.
52. Dunnill, P. M.; Fowden, L. Phytochemistry 1965, 4, 933-944.
53. Fang, S-T.; Li, L-C.; Niu, C-I.; Ts'eng, K-T. Sci. Sinica 1961, 10, 845-851.
54. Food Agric. Organ./World Health Organ. Expert Group Food Agric. Organ. Meetings Report Series 1965, 37, 1-71.
55. Sawaya, W. N. Unpublished results.
56. Tu, M.; Eustace, W. D.; Deyoe, C. W. Cereal Chem. 1978, 55, 766-772.
57. Bolley, D. S.; McCormack, R. H. J. Am. Oil Chem. Soc. 1951, 29, 470-472.

58. Feyt-Brugie, J.; Auriol, P.; Touze, A. C. R. Acad. Sci., Paris,
     Ser. D. 1968, 266, 2419-21.
59. Porterfield, W. M. Econ. Bot. 1955, 9, 211-223.
60. Whitaker, T. W.; Davis, G. N. "Cucurbits--Botany, Cultivation,
     Utilization"; Interscience Publishers: New York, NY, 1962.
61. Earle, F. R.; Jones, Q. Econ. Bot. 1962, 16, 221-250.

RECEIVED January 24, 1986

Protein–Nitrogen Conservation in Fresh Stored *Dioscorea* Yams

Godson O. Osuji[1], Robert L. Ory, and Elena E. Graves

Southern Regional Research Center, Agricultural Research Service, U.S. Department of Agriculture, New Orleans, LA 70179

Fresh and stored yams were examined for protein and selected nitrogen-metabolizing enzymes. Yams stored at room temperature lost weight (water), protein, and non-protein nitrogen. Sprouting depleted all of the protein. The major protein of yams is a 60-70% ethanol-soluble prolamine, virtually insoluble in NaCl or water. Purine-metabolizing enzymes in 7 Dioscorea yam cultivars were examined. Three of these (adenosine deaminase, uricase, allantoicase) give rise to ammonia; the others do not (xanthine oxidase, allantoinase). Allantoinase activity is low in yam tubers and may result in accumulation of allantoin. Yams also contain an alkaline proteinase that is active on yam prolamines, has a pH optimum at 9.5-10.0, and is stable to heat (60°C for 5 hours). It is not inhibited by thiol reagents, Ca, Mn, Mg, Zn or Cu but is strongly inhibited by Fe^3 and Fe^2. Dehydrated yam flakes were prepared from fresh yams in over 60% yield, as a potential means of storing yams safely for food uses in the tropics.

The yam tuber (Dioscorea spp.) is a primary carbohydrate staple crop in West Africa but it suffers considerable losses during postharvest storage due to fungal decay and/or sprouting (1). Since the yam is also a minor source of dietary nitrogen for many people whose diets contain little or no animal protein, loss of nitrogen by the tubers can be serious since other dietary proteins are not available. The major sources of nonprotein nitrogen in yams are the ureide compounds of the purine metabolic pathway (2). Yams are essentially tropical crops that cannot tolerate any frost or temperatures below 20°C (68°F) but the rate of growth increases with temperatures of 25-30°C (77-86°F), which are normal in areas where yams are grown (3, 4). The West African yam zone is the most important yam producing area of the world. About half of the world crop appears to be produced in Nigeria. Of 12.1 million metric tons produced in all of West Africa, about 9 million were produced in Nigeria (1). There are many species of Dioscorea of economic interest, of which ten are most important as

[1]Current address: Department of Agricultural Biochemistry, Anambra State University of Technology, Enugu, Nigeria.

sources of food. Of these, D. alata, D. cayenensis and D. rotundata are the major species grown in Nigeria and are the species examined in these investigations. Of these three, D. rotundata is by far the major species grown as a food crop, followed by D. alata and D. cayenensis.

In an earlier report (5), the decay of healthy yam tubers during storage was shown to be a result of catabolism of its proteins by an active α-glutamyl transpeptidase. There is also some alkaline proteolytic activity in the yam tuber (6), but little information is available on individual enzymes of the purine degradative pathway and on the properties of an alkaline proteinase that may function in yams during storage. This report describes the interrelation of five enzymes of ureide metabolism in fresh and stored yams, the release of ammonia in vitro by three of the enzymes that may provide an environment for alkaline proteinase activity in vivo, and the in vitro properties of an alkaline proteinase isolated from fresh yams. Using a method developed for the preparation of dehydrated flakes from sweet potatoes (7), dehydrated flakes were prepared from fresh yams. These flakes appear to be a suitable means for inactivating the yam enzymes and preserving the dietary nitrogen in yams for long term storage under tropical conditions.

Materials and Methods

Yam tubers of Dioscorea alata (Umudike cultivar), D. rotundata (asukwu and obiaturugo cultivars), and D. cayenensis (water yam and Nkokpu cultivars) were obtained from the National Root Crops Research Institute, Umudike, Nigeria. Some tubers were stored 6 or 12 months at room temperature (25-27°C), some in vacuum dessicators over a suitable dessicant, and some in paper bags placed in a dark cabinet (absence of circulating air). Fresh tubers were peeled by carefully scraping away the cork layer to minimize loss of outer tissue since much of the protein is concentrated here (8). They were then cut into 2 cu. cm. pieces, quickly frozen with solid CO_2 in 50 g portions in plastic bags, and stored in a freezer until needed.

Homogenization of Tuber Tissue. Frozen pieces of tissue (50g) were transferred to a Waring Blendor[3] and homogenized in 100 ml ice-cold 0.1 M K_2HPO_4 (4°C) and 0.1 ml β-mercaptoethanol at low speed for 3 min. The homogenate was filtered through two layers of cheesecloth and the filtrate centrifuged at 20,000 g. for 10 min. The pellet was discarded and the supernatant liquid was dialyzed against three changes of deionized water for 24 hr to remove low molecular weight sugars that might interfere with the phenylhydrazine assays for allantoinase and allantoicase activity.

Enzyme Assays. Adenosine deaminase (E.C. 3.5.4.4) was assayed by

---

[3] Trade names are given solely for the purpose of providing specific information. Their mention does not imply recommendation or endorsement by the U. S. Department of Agriculture over others not mentioned.

the method of Coddington (9) and absorbance was read at 265 nm in
a spectrophotometer over the first 5 min of reaction time. The
change in absorbance was used to calculate enzyme activity.
Xanthine oxidase (E.C. 1.2.3.2) was measured according to Bray
(10) with xanthine as substrate and oxygen as the electron
acceptor. The change in absorbance at 293 nm between 5 and 10 min
of reaction was used to calculate activity. Uricase (E.C.
1.7.3.3) was assayed according to Mahler (11) with oxygen as
electron acceptor. The change in absorbance was measured at 293
nm in a spectrophotometer over the first 5 min and used to
calculate activity. Allantoinase (E.C. 3.5.2.5) was measured with
allantoin as substrate by a modification of the methods of Singh
et al. (12) and Trijbels and Vogels (13). To 1 ml of allantoin
solution (3 mM in 0.1 M tris buffer, pH 7.4), 0.5 ml of tuber
extract was added and the mixture incubated at 40°C in a water
bath for 1 hr. The reaction was stopped by adding 1 ml each of
conc. HCl and phenylhydrazine (100mg/30ml deionized water) which
reacts with the glyoxylic acid formed from allantoin. Reaction
tubes were placed in boiling water for 5 min, then cooled rapidly
by immersing in an ice/water bath. After returning the tubes to
room temperature, 1 ml. of potassium ferricyanide (500 mg/30 ml.
deionized water) was added, mixed well, then centrifuged at 10,000
g for 5 min to remove precipitated proteins. Absorbance was
measured at 525 nm in a spectrophotometer. Controls contained
allantoin but no tuber extract and were treated the same as test
samples. Absorbance of controls was subtracted from test samples
and calculated from a standard curve prepared with glyoxylic acid.
        Allantoicase (E.C. 3.5.3.4) was assayed with allantoic acid
as substrate by a modification of the methods of Trijbels and
Vogels (13) and Ory, et al. (14). To 0.5 ml allantoic acid
solution (3.5 nM in 0.1 M tris buffer, pH 7.4), 0.5 ml of the
tuber extract was added and incubated at 40°C in a water bath for
1 hr. Test tubes were placed in ice, then 1 ml each of conc. HCl
and phenylhydrazine (100 mg/30 ml deionized water) were added,
followed by 1 ml of the potassium ferricyanide solution and mixing
to promote reaction with glyoxylic acid. The precipitate was
removed by centrifugation (as above) and absorbance read at 525 nm
in the spectrophotometer. Controls containing allantoic acid
substrate but no tuber extract were treated as above and
absorbance of controls was subtracted from test samples to
calculate activity. Ureidoglycine and ureidoglycolate produced
from allantoic acid by tuber extracts were determined by the
differential glyoxylate method of Trijbels and Vogels (13) after
the tuber extracts and allantoic acid had been incubated as
described for allantoicase.

Proteins and Proteinase Isolation/Assay. Pieces of tuber were
homogenized (100g/150 ml absolute ethanol) in a Waring Blendor for
3 min (as before), filtered through cheesecloth, and the filtrate
centrifuged at 10,000 g for 15 min. The supernatant was made to
10% with trichloroacetic acid to precipitate the protein and
stored at 0°C for several hours to separate the yam prolamine.
After centrifuging at 10,000 g for 15 min, the precipitate was
suspended in a minimum volume of water and dialyzed against

deionized water overnight, then freeze-dried and stored at 5°C until needed.

For the proteinase, 100 g of yam cubes was homogenized in 150 ml ice cold (4°C) deionized water containing 2 g polyvinylpyrrolidone (PVP) and 100 mg ethylenediaminetetraacetic acid (EDTA) in a Waring Blendor at low speed for 3 min, filtered through two layers of cheesecloth, then centrifuged as before. The supernatant liquid was made up to 85% saturation with solid $(NH_4)_2$ SO_4; the precipitate was removed by centrifuging at 10,000 g for 15 min, then dissolved in 2-5 ml of deionized water for dialysis against deionized water overnight. The dialyzed enzyme was freeze-dried for storage in the freezer until needed. In some cases, the original homogenate was made up to 35% with $(NH_4)_2$ SO_4 to remove small amounts of protein; then the filtrate was raised from 35% to 85% with $(NH_4)_2$ SO_4, precipitated, centrifuged, dialyzed, and freeze-dried as before, to effect a higher purification.

Protein contents of tuber extracts were determined by the method of Lowry et al. (15) using bovine serum albumin as the standard.

For activity assays, proteinase solutions were made fresh daily (10 mg freeze-dried solids in 1 ml pH 10.0 phosphate buffer, 0.1 M). Two ml of the proteinase, 0.5 ml of substrate (azocasein or other proteins in pH 10 buffer), 0.3 ml of 0.1% EDTA, and deionized water were made up to a volume of 3.5 ml. Reaction tubes were incubated in a 40°C water bath for 1 hr, then the reaction was stopped by addition of 1 ml 5% TCA. After removal of the precipitated proteins by centrifugation, absorbance was read at 366 nm (for azocasein) or by the Lowry method (15) for other substrates, to calculate proteinase activity with the different substrates.

For electrophoresis of the proteinase, saturated solutions of crude and purified preparations were made in 0.1 M Tris-glycine buffer, pH 10.5, and electrophoresed in 7% polyacrylamide gel with and without sodium dodecyl sulfate (SDS).

Results and Discussion

A summary of the proximate analyses of these yam tubers is shown in Table 1. On a fresh weight basis, protein is quite low because of the high water and starch contents of yams (dry weight protein content is 4-8%). Yams are rarely consumed alone but, depending upon the economic status and tastes of the consumer, they are accompanied by meat or fish, green vegetables, or spices. Where animal protein is not included, conservation of the limited amounts of crude protein in yams is essential for areas that depend upon yams as their primary food source.

Because yams are poor sources of dietary nitrogen, a knowledge of nonprotein nitrogen metabolism in yams would provide useful information for retaining the amount present in fresh tubers and preventing the usual losses suffered during storage in the tropics. Yams accumulate large quantities of nonprotein nitrogen as allantoin (16) but the ureide compounds (allantoin, allantoic acid, etc.) are more important intermediates in the nitrogen-rich plant species (17-19) than in nitrogen-deficient

Table I. Summary of Proximate Analyses of Fresh Yam Tubers [1].

Component	Species		
	D. alata	D. cayenensis	D.rotundata
	%	%	%
Moisture	65-73	83	58-73
Carbohydrate	22-29	15	23
Fat	0.03-0.27	0.05	0.12
Crude Protein (N x 6.25)	1.1-2.8	1.0	1-2
Crude Fiber	0.6-1.4	0.4	0.3-0.8
Ash	0.7-2.1	0.5	0.7-2.6

[1] Data taken from Coursey ([1]), pp. 154-157.

root tubers. However, ureides are known to be metabolized by plant roots ([20-22]).

Five enzymes of the purine degradative pathway in two species of yams were investigated to obtain a better understanding of nitrogen metabolism in fresh and stored tubers. The results (Table 2) show that 12 months storage had varied but striking effects on activity of adenosine deaminase, xanthine oxidase, uricase, allantoinase, and allantoicase in both D. alata and D. rotundata. In the general scheme of purine degradation, adenosine is deaminated to yield xanthine, which in turn is oxidized to uric acid, the substrate for uricase. Subsequent deamination by uricase produces allantoin, the substrate for allantoinase. Allantoin degradation to allantoic acid provides the substrate for allantoicase, which converts this to urea and gloxylic acid. Thus, low activity (or inhibition) of the last two enzymes of this chain could be responsible for the accumulation of allantoin in certain plants. Whereas the water yam cultivar shows a general increase in activity for all five enzymes during storage, the other cultivars show mixed effects; some lose activity and some increase activity upon storage. Comparing allantoinase/ allantoicase activities, these enzymes in the D. alata cultivars increased during storage whereas those in the D. rotundata cultivars decreased. Such decreased activities during storage may provide a mechanism for increased accumulation of allantoin/ allantoic acid and subsequent retention of nonprotein nitrogen in the tubers. This also provides useful information for breeding yam tubers with higher nitrogen; tubers with lower activities of these enzymes would be more desirable. If the specific activities of these two enzymes in fresh yams are compared (Table 3), it appears that allantoicase activity is higher in all cultivars. This suggests that allantoinase may be the limiting enzyme in the purine degradative pathway.

Table 2. Effects of 12 Months Storage on Relative Activities of Ureide Enzymes in Yam Tubers.

Species/Cultivar	Adenosine Deaminase		Xanthine Oxidase		Uricase		Allantoinase		Allantoicase	
	Fresh %	Stored %	Fresh %	Stored %	Fresh %	Stored %	Fresh %	Stored %	Fresh %	Stored %
D. alata (umudike)	100	46	100	16	100	120	100	118	100	100
D. cayenensis (water yam)	100	155	100	210	100	152	100	116	100	112
D. rotundata (obiaturugo)	100	89	--*	100	100	109	100	97	100	82
D. rotundata (asukwu)	100	159	100	225	100	48	100	45	100	37

Table 3.   Specific   Activities   (μM/min/mg)   of   Allantoinase   and Allantoicase in Fresh Yams.

Species/Cultivar	Allantoinase	Allantoicase
D. alata (umudike)	1.1	1.5
D. cayenensis (water yam)	0.8	2.5
D.rotundata (obiaturugo	1.0	1.7
D.rotundata (asukwu)	3.6	11.8

Release of Ammonia.   Of the five enzymes, three lead to release of amide/amino   nitrogen   as   ammonia   during   ureide   metabolism: adenosine   deaminase,   uricase,   and   allantoicase.   This   could provide a mechanism for the overall net loss of nitrogen by the tubers or for accumulation in the tissues.
Accumulation of NH_3 is believed to result in the build up of ureides in plants (23). Since yams are traditionally stored in open barns in West Africa for several months, both protein and nonprotein nitrogen can be lost. Yams are the primary source of dietary carbohydrate but are also a minor source of nitrogen. Kjeldahl analysis of yam tissue before and after storage showed that D. alata Umudike cultivar lost 31% of its total nitrogen after 12 month's storage and 65% of the nonprotein nitrogen. D. rotundata asukwu cultivar lost 15% of the total nitrogen and 50% of the nonprotein nitrogen after storage. Loss of ammonia by purine degradation could be significant since the tubers are already low in nitrogen.

Alkaline Proteinase Activity in Yams.   The release of ammonia at several stages during ureide metabolism suggested a potential for alkaline conditions in yam tubers, rather than the usual neutral or acid conditions generally found in seeds and plants.
Because yams are stored in open systems at ambient temperatures (usually warm), tuber tissue was examined for proteinase activity at 40°C. Some tubers had high apparent polyphenoloxidase activity upon peeling of the tubers (tissue turned deep purple at the peeled surface) so that PVP was added to extracts to combine with polyphenolic compounds and protect the proteinase from reacting with these compounds. Earlier studies had shown some inhibition of alkaline proteinase activity by ferric ion (24) so that EDTA was also added to the extracts to chelate any free iron. Two alkaline pH optima were found, at 9.0 and 10.5. The alkaline proteinases of white potatoes (Solanum tuberosum) have pH optima between 8.6 and 9 (25) and those of Carilla chocola tubers have pH optima between 8.0 and 9.5 (26,27), suggesting that alkaline proteinases of Dioscorea tubers may have the highest pH optima yet reported in plants.
     The temperature optimum for this proteinase may also be a result of evolutionary adaptation to the hot temperatures experienced during growing and postharvest storage in West Africa.

Maximum activity was found at 60° (24). As seen in Fig. 1, the enzyme lost only 10-15% of its activity at 60°C after 5 hr at that temperature, suggesting a remarkable stability toward heat. However, all tests were conducted at 40°C to avoid denaturation effects on proteinase substrates and difficulties involved with maintaining the higher temperature. Optimal temperature of the thiol proteinase of white potato is 40°C (25) and a membrane-bound proteinase in germinating pea seeds is stable at 60°C for 1 hr. (28). The yam tuber proteinase does not seem to be membrane-bound, since it is easily extracted from the tissue without detergents. To study substrate specificity, two assays were used. In initial tests (pH optimun, heat stability/optimum), azocasein was employed as a chromogenic substrate because of its sensitivity. Because the yam proteinase had only half or less of the activity of trypsin and pepsin on azocasein, additional tests on substrates employed the Lowry method (15) to measure hydrolysis of substrates by yam proteinase. Results of these tests are summarized in Table 4. Lowest activity was measured with azocasein, peanut globulin (arachin), egg albumin, bovine serum albumin and hemoglobin, and β-lactoglobulin, all readily soluble proteins. Highest activity was exhibited towards the two prolamines, wheat gliadin and yam protein; both proteins extracted with ethanol. The yam proteinase showed as much activity toward the yam prolamine as it did on wheat gliadin. This suggests that the yam proteinase may function in vivo by hydrolyzing the difficultly-soluble prolamines during long term storage of the tubers under tropical conditions. Results in Table 4 also suggest that the proteinase is not a serine proteinase. Tests with the two effectors, leupeptin and pepstatin-A, showed no inhibition of activity with azocasein as substrate. The observed inhibition of activity by ferric ions suggests that it is not a metalloenzyme. Thus, it appears to be different from the known plant proteinases.

Attempts to purify the enzyme by Sephadex chromatography were not successful because of excessive losses of activity after passage over the columns. Such losses after chromatography suggested that the proteinase may be dissociating into subunits (or changing conformation) that have very little activity. The final bands of activity were so faint that they could not be photographed. The purified enzyme gave a single, very light band upon electrophoresis and attempts to improve sharpness of the band in the gels were not successful. Electrophoresis with molecular weight markers provided a crude estimate of the size. The crude enzyme (undissociated) did not enter the gel as readily as did rabbit muscle myosin (205,000), but with sodium dodecyl sulfate (SDS), the subunits migrated faster than did the carbonic anhydrase marker (29,000).

Protein in Stored Tubers. The heat tolerance, alkaline pH optimum, and substrate specificity of the yam proteinase suggested this enzyme as the primary cause of protein loss during storage. Fig. 2 shows fresh yams less than a month after harvest and a similar tuber after 4 months storage at ambient temperature (24-27°C) in open paper bags in a closed laboratory cabinet. The temperature, humidity, and limited air flow promoted sprouting of the tubers. The new sprout was over 25 cm. in length and attempts

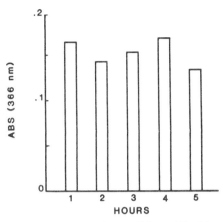

Figure 1.   Heat Stability of Yam Tuber Alkaline Proteinase
at 60°(with azocasein as substrate).

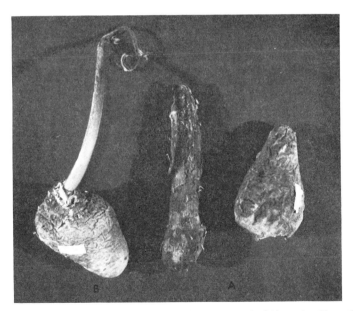

Figure 2.   Photograph of Yam Tubers, Fresh (A) and after 4
Months Storage (B) Showing a Typical Sprout
that Occurs in Stored Tubers.

Table 4. Substrate Specificity of Yam Tuber Alkaline Proteinase
(Activity: mg protein hydrolyzed)/10 mg enzyme/2hr).

Substrate	Azocasein	Azocasein + Leupeptin	Azocasein + Pepstatin-A
Activity (at 366 nm)	0.018	0.018	0.016

Substrate	Peanut Arachin	Egg Albumin	Bovine Serum Albumin	Bovine Hemoglobin	β-Lacto-globulin	Wheat Gliadin	Yam Protein
Activity (at 750nm)	0.260	0.390	0.305	0.325	0.265	0.771-0.805	0.860

to extract protein from the residual tuber were fruitless, indicating a rapid and severe loss of protein nitrogen. Since the tubers also lose nitrogen via ureide metabolism during storage, consumption of yams after extended storage under adverse conditions could introduce nutritional disorders where other sources of protein are lacking. Either the tubers would have to be consumed soon after harvest or a method must be found to improve the traditional ways of storing yams in the tropics.

An attempt was made to prepare dehydrated flakes from yam tubers by the SRRC-developed method for preparing sweet potato flakes (7). This would inhibit all enzyme activities, remove the large amounts of water (Table 1), and retain the initial protein and nitrogen of the tubers. Since the SRRC process is energy-intensive and applications in West Africa would have to depend primarily on available manpower, 12.5 kg of yam tubers were peeled by hand, diced into 2 x 2 cm. pieces, boiled in water and stirred with a wooden paddle in a large kettle, then poured into the mechanical filter for removing fibrous material and mashing. The puree was manually transferred to a hopper and drum-dried by passing steam through the slowly revolving drum (the only step that will require fossil energy in Africa). The flakes obtained (Fig. 3) were off-white in color, had good texture and represented over 60% recovery on a dry weight basis. Over 95% of the original nitrogen of fresh tubers was recovered, suggesting a potential means of preserving yams without fear of losses due to sprouting, insects, fungal degradation, etc., in the tropics. Yam flakes should be more stable for extended storage than fresh yams and, as flakes, they would require less storage space.

Figure 3.   Dehydrated Flakes Prepared from Fresh Peeled Yam Tubers by the SRRC-Sweet Potato Flaking Process.

Literature Cited

1. Coursey, D. G.: "Yams"; Longmans, Green: London, 1967; p. 172; 135.
2. Osuji, G.O.; Ory, R. L. J. Agric. Food Chem., In Journal review 1985.
3. Copeland, E. B. Philipp. J. Sci. 1916, 11, 277.
4. Prain, Sir D.; Burkill, I. H. Ann. R. Bot. Gdn., Calcutta 1936, 14, 1.
5. Osuji, G. O. Acta Biol. Med. Germ., 1981, 40, 1497.
6. Osuji, G. O.; Umezurike, G. "The Biochemistry and Technology of the Yam Tuber"; ASUTECH Press, Enugu, Nigeria, 1985.
7. Wadsworth, J. I.; Koltun, S. P.; Gallo, A. S.; Ziegler, G. M.; Spadaro, J. J. Food Technol. 1966, 20 (6), 111.
8. Walter, W. M.; Collins, W.; Purcell, A. E. J. Agric. Fd. Chem. 1984, 34, 695.
9. Coddington, A. Biochim. Biophys. Acta 1965, 99, 442.
10. Bray, R. C. "The Enzymes." Eds. : Boyer, P. D.; Lardy, H.; Myrbäch,K., Academic Press, Inc., New York, 1963; no. 7, 533.
11. Mahler, H. R. "The Enzymes." Eds. : Boyer, P. D.; Lardy, H.: Myrbäch, K., Academic Press, Inc., New York, 1963; vol. 8, 285.
12. Singh, R.; St. Angelo, A. J.; Neucere, N. J. Phytochemistry 1970, 9, 1535.
13. Trijbels, F.; Vogels, G. D. Biochim. Biophys. Acta 1966, 113, 292.
14. Ory, R. L.; Gordon, C. V.; and Singh, R. Phytochemistry 1969, 8, 401.
15. Lowry, O. H.; Rosebrough, N. J.; Farr, A. L.; Randall, R. J. J. Biol. Chem. 1951, 193, 265.
16. Ueda, H.; Sasaki, T. J. Pharm. Soc. Japan 1956, 76, 745.
17. Fosse, R., C. R. Acad. Sci. 1926, 182, 869.
18. Mothes, K.; Engelbrecht, L. Flora 1952, 139, 586.
19. Krupk, R. M.; Towers, G. H. N. Canad. J. Bot. 1959, 37, 539.
20. Brunel, A.; Capelle, G. Bull. Soc. Chim. 1947, 29, 427.
21. Kushizaki, M.; Ishiguka, J.; Alkamatsu, F J. Sci. Soil Manure, Japan 1964, 35, 232.
22. Ishizuka, J.; Okino, F.; and Hoshi, S. J. Sci. Soil Manure, Japan 1970, 44, 78.
23. Thomas, R. J.; Feller, U.; Erismann, K. H. Plant Physiol. 1979, 63, Suppl. 50.
24. Osuji, G. O.; Ory, R. L.; Graves, E. E. J. Agric. Food Chem., in journal review 1985.
25. Santarius, K., and Belitz, H. D. Planta 1978, 141, 145.
26. Ryan, C. A. Ann. Rev. Plant Physiol. 1973, 24, 173.
27. Tookey, H. L.; Gentry, H. S. Phytochemistry 1969, 8, 989.
28. Ashton, F. M. Ann. Rev. Plant Physiol. 1976, 27, 95.

RECEIVED January 24, 1986

INDEXES

Author Index

Subject Index

Production and indexing by Karen McCeney
Jacket design by Pamela Lewis

Elements typeset by Hot Type Ltd., Washington, DC
Printed and bound by Maple Press Co., York, PA

RECENT ACS BOOKS

"Phenomena in Mixed Surfactant Systems"
Edited by John F. Scamehorn
ACS Symposium Series 311; 356 pp; ISBN 0-8412-0975-8

"Chemistry and Function of Pectins"
Edited by Marshall Fishman and Joseph Jen
ACS Symposium Series 310; 286 pp; ISBN 0-8412-0974-X

"Fundamentals and Applications of Chemical Sensors"
Edited by Dennis Schuetzle and Robert Hammerle
ACS Symposium Series 309; 398 pp; ISBN 0-8412-0973-1

"Polymeric Reagents and Catalysts"
Edited by Warren T. Ford
ACS Symposium Series 308; 296 pp; ISBN 0-8412-0972-3

"Excited States and Reactive Intermediates:
Photochemistry, Photophysics, and Electrochemistry"
Edited by A. B. P. Lever
ACS Symposium Series 307; 288 pp; ISBN 0-8412-0971-5

"Artificial Intelligence Applications in Chemistry"
Edited by Bruce A. Hohne and Thomas Pierce
ACS Symposium Series 306; 408 pp; ISBN 0-8412-0966-9

"Organic Marine Geochemistry"
Edited by Mary L. Sohn
ACS Symposium Series 305; 440 pp; ISBN 0-8412-0965-0

"Fungicide Chemistry: Advances and Practical
Applications"
Edited by Maurice B. Green and Douglas A. Spilker
ACS Symposium Series 304; 184 pp; ISBN 0-8412-0963-4

"Petroleum-Derived Carbons"
Edited by John D. Bacha, John W. Newman and
J. L. White
ACS Symposium Series 303; 416 pp; ISBN 0-8412-0964-2

"Historic Textile and Paper Materials: Conservation
and Characterization"
Edited by Howard L. Needles and S. Haig Zeronian
Advances in Chemistry Series 212; 464 pp; ISBN 0-8412-0900-6

"Multicomponent Polymer Materials"
Edited by D. R. Paul and L. H. Sperling
Advances in Chemistry Series 211; 354 pp; ISBN 0-8412-0899-9

For further information contact:
American Chemical Society, Sales Office
1155 16th Street NW, Washington, DC 20036
Telephone 800-424-6747